Telephone Installation Handbook

Dedication

To the memory of Mike Thompson G4HEU

Friend, colleague, and fellow engineer of the old school.

Telephone Installation Handbook

Second edition

Steve Roberts IEng. MIIE(elec)

Newnes

OXFORD   AUCKLAND   BOSTON   JOHANNESBURG   MELBOURNE   NEW DEHLI

Newnes
An imprint of Butterworth-Heinemann
Linacre House, Jordan Hill, Oxford OX2 8DP
225 Wildwood Avenue, Woburn, MA 01801-2041
A division of Reed Educational and Professional Publishing Ltd

 A member of the Reed Elsevier plc group

First published 1997
Reprinted 1999, 2000
Second edition 2001

British Library Cataloguing in Publication Data
A catalogue record for this book is available from the British Library

ISBN 0 7506 5269 1

Typeset by David Gregson Associates, Beccles, Suffolk
Printed in Great Britain by Biddles Ltd, www.Biddles.co.uk

Contents

Preface

Current telephone legislation

Up until 1 January 1985, British Telecom held the monopoly for the supply of the primary telephone instrument to be connected to the Public Switched Telephone Network (PSTN) in the United Kingdom. (Such networks are currently owned and operated by British Telecommunications plc, Kingston Communications (Hull) plc, Mercury Communications Ltd, and various cable companies.) Government legislation that came into force on that date opened the way for companies in the private sector to supply approved telephones for connection to domestic direct exchange lines.

A further relaxation, introduced on 1 December 1986, allowed users, or their agents, to install extension wiring and telephone sockets and to connect them to the PSTN. However, in order to do this, the user must have a master socket installed by the appropriate network operator.

To the general public, this change of policy meant that the telephone apparatus could be purchased outright, in some cases for less than £10, instead of paying a rental charge of around £12 a year. It also opened the way for enterprising individuals, or other people, to install extensions of their own.

Rules and regulations

Just as there are requirements when installing mains wiring, there are a few regulations governing the installation of telephone wiring. These regulations are intended to prevent both damage to existing equipment and injury to personnel using it. Armed with the knowledge of these regulations, plus a few tools and the necessary materials, most people should be capable of planning and implementing an installation.

Supported by an extensive glossary of technical terms, abbreviations and acronyms, this book is aimed at the inexperienced. It is intended to strip away the mystique surrounding telephone wiring, to give information and guidance, to help in the selection of tools, materials, and telephone apparatus, and to help the reader avoid some possible pitfalls. Above all, it is meant to be a practical book. It is for those who set out, within the private sector, to install extensions and systems for others; for managers in industry, to enable them to communicate effectively with their own installers and maintainers; and for those who prefer to do their own work around the house. While many new types of apparatus and equipment will continue to come to market, the fundamental techniques of connecting them, set out in this book, will be of lasting value.

Second edition

Since the first edition of this book was published, advancements have been made in a number of areas, most notably in the more widespread use of ISDN and of structured cabling. Additional material has therefore been included to cover these topics, and the opportunity has been taken to clarify, enhance, or update some sections where the need has arisen. However, the purpose of the book remains unchanged – to be a practical guide to installation.

Steve Roberts

Acknowledgements

The author would like to thank all those who have been instrumental in the preparation of this book and, in particular, to John Weller, of Rapid Lines Telecommunications, for endless assistance in all sorts of ways; to my wife Deryn for keeping the computer under control; to my mother 'Gary', for providing the historical sketches; to Margaret Salmon for lending Internet and other assistance; to Eddie Ruff for providing an electrician's perspective; to Messrs Comtec Telcom Products for supplying much useful information; and to Henry Boettinger for his advice and encouragement in the early stages.

1
The telephone system

Each subscriber's line consists of a single pair of wires, connecting the subscriber's premises to the local exchange. Such a line is called a 'direct exchange', line or 'DEL'. (In American terminology, the exchange is called the central office, abbreviated to 'CO', and a direct exchange line is called a 'CO line'.)

This simple connection, however, is used for a remarkable number of functions. Firstly, the line carries the voice frequency (VF) communication, in both directions. Such communication may be in the form of speech, of modem or facsimile (fax) tones, or of other tones. Secondly, the line also carries signalling information, both from the subscriber to the exchange and vice versa. Some signalling is in direct current (DC) form, and some is in alternating current (AC) form.

Let us consider a telephone call from initiation, and observe how the progress of the call is marked by various forms of signalling. Figure 1.1 shows the essential parts of a crude telephone and exchange line connection, which may aid understanding.

Stage 1. The subscriber lifts the handset, which releases a switch mechanism in the telephone called a hookswitch. (The name recalls the days when the earpiece ('receiver') was suspended from a hook when not in use, this hook being arranged to operate a switch. The term 'hang up' has the same origin.) This has the effect of connecting some part of the telephone's circuitry across the ends of the two wires of the exchange line, providing a DC path. At the exchange, a 48 volt (V) power supply is indirectly connected to each subscriber's line. When the DC path is completed, current will flow, and the 'local loop' is formed. Providing the current flow is greater than about 20 milliamperes (mA), the line relay in the exchange, whose

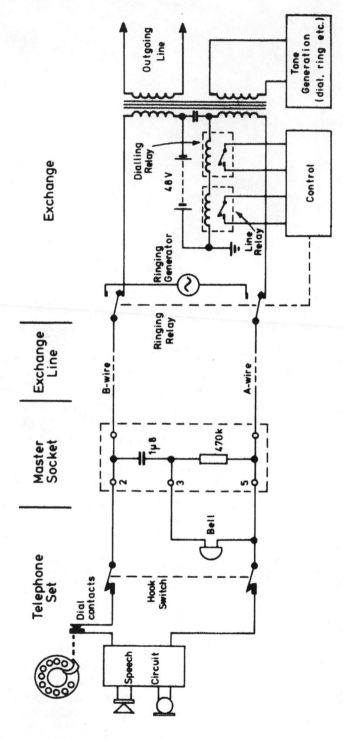

Figure 1.1 *Crude telephone with exchange line connection.*

coil is connected in series with the local loop, will operate, thus signalling to the exchange that service is required.

Note that the description of the operation here and in ensuing paragraphs refers to the use of relays. In specific exchanges, these may well be replaced by solid-state sensors and switches, but this does not affect the theory of operation.

Stage 2.   The exchange now connects the subscriber's line to the system and indicates that it has done so by applying the familiar 'dialling tone' to the line.

Stage 3.   The subscriber now indicates to the exchange the number of the called party. This may be achieved in one of two ways:

● By means of loop disconnect (LD) or 'pulse' dialling – turning the current on and off.
● By means of tone (dual tone multi-frequency – DTMF) – dialling – sending 'musical' tones for the number dialled, without interrupting the connection.

Loop disconnect dialling has been in general use since the early days of telephony. It operates by temporarily disconnecting the local loop – once for number one, twice for number two, etc., and ten times for number zero. The temporary disconnection is too short for the line relay in the exchange to drop out, since these relays are designed to be of sluggish action. However, another relay in the local loop has a faster reaction time, and it operates in sympathy with the makes and breaks of the dialling contact. While dialling is in progress, the impulses are prevented from reaching the earpiece by closure of a muting switch. Obviously, for the dialling to be successful, the timing of the pulses in respect of both the pulse width and the time between successive pulses, as well as the time between successive digits, has to be controlled. Figure 1.2 shows the timing of these pulses as applicable to the UK telephone network.

(Loop disconnect dialling is a development of the first form of automatic dialling to be used. It was patented in 1891 by Almon Strowger, a funeral director of Kansas City. He suspected that he was losing business to a rival because calls for 'a funeral director' were being manually routed by the telephone operator to his rival. Legend has it that the telephone operator was the rival's sister!)

In the case of tone dialling, which can only be accepted by the more modern telephone exchanges (such as System X), each digit is represented by two simultaneous tones. These tones are generated within the telephone, and selected according to which button of the keypad is pressed. Figure 1.3 shows the keypad layout and the corresponding combination of tones that will be transmitted along the telephone line to the exchange.

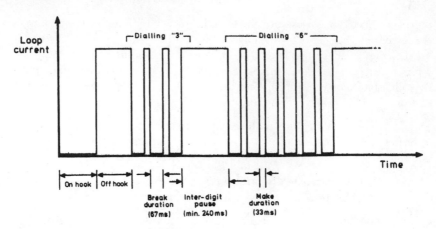

Figure 1.2   *Loop disconnect dialling – pulse timing.*

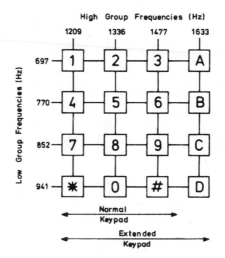

Figure 1.3   *MF dialling – keypad layout and tone frequencies.*

The tones are decoded at the exchange and the call routing controlled accordingly. Tone, or DTMF, dialling is both faster and more reliable than the older loop-disconnect method.

- It is faster because each digit takes just a few tens of milliseconds (ms) to transmit – the speed is dictated largely by the dexterity of the operator in pressing the buttons. However, pulse-dialled digits will take between 1 and 2 seconds (s) each.
- It is more reliable because the decoding and switching operations are

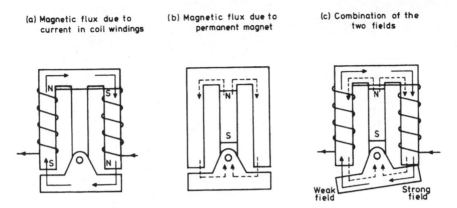

(a) Magnetic flux due to current in coil windings

(b) Magnetic flux due to permanent magnet

(c) Combination of the two fields

Figure 1.4   *Telephone bell – principle of operation.*

accomplished electronically, so sensitive electromechanical devices can be eliminated.

- There is another benefit to the tone-dialling system. The tones may be used for signalling purposes after the connection has been made. This permits, for instance, certain answering machines to give up their stored messages in response to a predetermined code. It has also allowed BT to introduce their 'Star Services', whereby subscribers may store frequently used telephone numbers at the exchange, set up conference calls, and control call diversion, call barring, and other facilities.

Stage 4.   The subscriber now hears one of a number of possible tone signals indicating number ringing, number engaged, or number unobtainable.

The distant exchange now applies a ringing signal to the line of the called party. It is perhaps worth remembering that at this stage of the call, a considerable amount of equipment is in use, but there is no revenue resulting, since the voice frequency path has not been established. It is therefore in the interest of the service provider that the attention of the called party is attracted as soon as possible. The bell, or other audible signal, needs to be as loud as possible, within acceptable limits.

The ringing signal (or 'call arrival indication') is an AC signal of around 90 V, at a frequency of about 20 Hertz (Hz). The telephone bell, which is energized by this signal, is shown diagrammatically in Figs 1.4 and 1.5. The bell consists of an armature that is pivoted between two electromagnets consisting of two coils wound onto permanent magnet cores. The two coils are wound so that the magnetization due to the ringing current is of opposite polarity on each side of the structure. A permanent magnet bias ensures that the armature will be attracted first to one side, then the other, in sympathy with the alternating magnetic field.

Note that the bell is powered through a capacitor, normally located in the master socket, so that the bell coil windings do not provide a DC path to

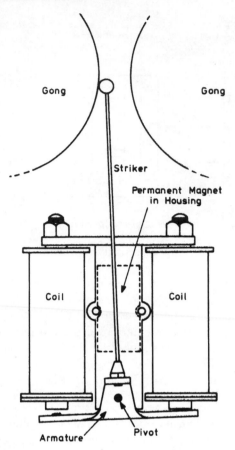

Figure 1.5   *Telephone bell – physical arrangement.*

complete the local loop. At the time when the ringing signal is applied to the line, only the bell circuitry is connected to the line, the rest of the telephone being disconnected by means of the hook switch.

When one of a number of parallel-connected telephones is being used for pulse-dialling, the continual interruptions of the line current may cause the bells of the other telephones to 'tinkle' in sympathy. In order to overcome this annoying phenomenon, a second pair of contacts in the hook switch is used to connect a low resistance across the bell circuit.

Electronic ringing circuits are becoming increasingly common in telephones today. There are a number of benefits associated with the use of such circuits.

- They are generally smaller and lighter than the conventional bell.
- Variation of both pitch and of volume is quite easy to arrange. Use of different pitches allows the identification of one of several telephones in an

office, and electronic volume control may allow the volume to increase the longer the phone rings.

In the conventional bell system, the physical movement of the gong is directly controlled by the amplitude and frequency of the ringing signal. In the case of the electronic ringing circuit, the signal is merely rectified and smoothed, and used as a power source for the ringer. One advantage of this approach is that it provides another method of overcoming the problem of bell tinkle.

Ringing circuits are commonly available as integrated circuits, and so elaborate tone generation is quite feasible – for instance, the low-frequency alternation between two different tones to produce a warbling effect.

The electrical signals still have to be converted to audible tones, though, and electromagnetic or piezoelectric transducers may be used for this purpose. When the telephone is answered by the called party, the hook switch is closed, thus completing the local loop. The exchange recognizes the off-hook condition, and responds by replacing the ringing signal with the voice frequency path from the calling end. This state is maintained until one party hangs up, and their local loop is broken.

It is a feature of the telephone system that the call path will be maintained until the calling party hangs up. This means that the called party can replace the receiver or unplug the telephone and move to a new location to take the call in comfort and/or privacy. It also means that charges for the call continue to accrue until the calling party hangs up. There have been occasions when a call has been made, and terminated properly by the recipient, but not by the call originator who inadvertently left the receiver off the hook. In one instance, a transatlantic call was made immediately prior to a two-week absence, and a very large charge indeed was incurred. Happily, in this case, BT waived the charge when advised of the circumstances.

The speech circuit

The speech circuit performs three basic functions:

- To balance the send and receive signal levels so that the correct talk and listen sensitivities are achieved.
- To provide a degree of 'sidetone' – i.e. the application of some of the microphone signal to the earpiece. If this were not done, the user would complain of a 'dead' earpiece.
- To provide regulation to compensate for the varying line lengths. The resistance of the local loop may vary between virtually zero and 1000 ohms (Ω) – for which the telephone must be able to compensate. This is classically achieved (in the 700-series telephone, for instance) by introducing a circuit whose resistance decreases with increasing DC line

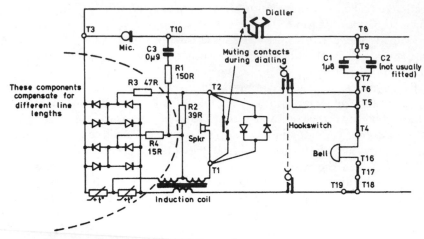

The induction coil,
R1, R2, C1 & C3 provide
level balancing and
sidetone. C1 is also
used as a bell capacitor.

Figure 1.6   *700-Series telephone circuit diagram.*

current, and using this to shunt the microphone. A similar arrangement, though with rather less loss, is used for the receive signal.

The BT 700-series telephone was the standard issue for many years and will be familiar to all. It was the first to be available in a variety of colours, and was equipped with an internal bell and rotary dial, although some push-button versions were produced. The circuit diagram of the basic 700-series telephone is reproduced in Fig. 1.6.

Elimination of induced noise

Some readers may be surprised that the telephone network could ever work, given the very long line lengths employed and the amount of electrical pollution, mostly mains hum, which must be picked up. The use of a balanced line and differential amplifiers solves this problem.

The principle is that the voice frequency signals are applied as a difference signal between the two wires of the pair, and at the distant end, only the difference signal is amplified. This of course assumes that the electrical characteristics of each leg are identical. Any imbalance due to a fault condition will cause hum and other noise to become apparent. Such faults may be caused by leakage paths to earth from one wire and not to the other (perhaps in turn due to dampness in a cable), or by complete disconnection

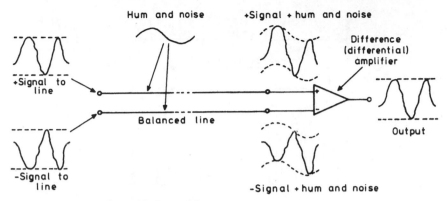

Figure 1.7 *Principle of balanced line.*

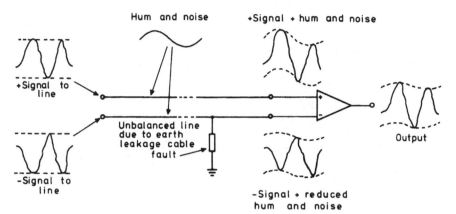

Figure 1.8 *Fault condition on balanced line.*

of one wire further down the line. Figs 1.7 and 1.8 illustrate the principle and possible problems that may be encountered.

Voice channel bandwidth

As any telephone user is aware, the quality of the transmission of the human voice over a telephone link is far from perfect. This is not a problem however. The human voice contains many more sounds than are actually required for intelligible speech, and savings may be made in the costs of transmission equipment if only the essential range of frequencies is transmitted. The voice bandwidth, as it is called, is limited to the range 300–3000 Hz (3 kHz), which is transmitted within a voice channel that is 4000 Hz (4 kHz) wide (see Fig. 1.9). The apparent wastage (the difference between the voice channel and the voice bandwidth) provides a guard band

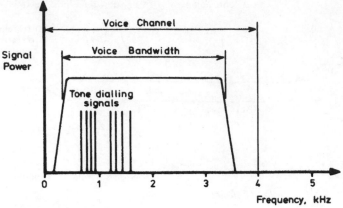

Figure 1.9   *Voice channel bandwidth.*

between adjacent channels when frequency division multiplexing (shortly to be described) is in use and also allows various system control signals to be inserted in the transmission path.

Beyond the local loop

The connection between each subscriber and the local exchange is normally made by means of a pair of wires, which will almost certainly be incorporated at some stage within a multipair cable carrying many subscribers' lines. Continuation of this method of distribution beyond the local exchange would attract a very high cost indeed due to the amount of copper in use, and the associated amplifiers, etc.

For this reason, methods have been devised of transmitting many channels at once through one physical, or radio, connection. In one such system, channels are formed and stacked one above the other in frequency to form a 'baseband'. The baseband is then transmitted by coaxial cable, by a radio link (usually in the microwave range) or by fibre-optic cable. This method is known as frequency division multiplex, or FDM.

There is another method of multiplexing, by means of time division. In its simplest form, each channel is sampled briefly in sequence, in the manner of a motor-driven rotary switch. At the distant end, a similar rotary switch is driven in synchronism with the first, and thus each sampled channel may be separated out again (see Fig. 1.10). Providing there are sufficient samples per unit time, the original signals may be reconstructed. Of course, the sampling and switching are performed electronically rather than mechanically, and modern systems convert the analogue signals to digital form first, which then provides the basis of the pulse code modulation (PCM) method of transmission.

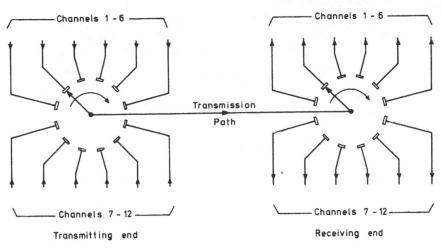

Figure 1.10 *Principle of time-division multiplex.*

By using fibre-optic links, where the transmission medium is light, very wide bandwidths with associated high traffic densities may be employed. Fibre-optic links have a high immunity to both interference and eavesdropping, and the cables employed are less bulky and require less protection from chemical contamination and moisture than their copper counterparts.

Using service providers other than BT

Since the liberalization of the telephone service, licences have been granted to other service providers such as Cable and Wireless. Immediately, the problem of fair competition arises, since BT have inherited an enormous infrastructure of telephone equipment and cabling, and other carriers would need to install duplicate systems before starting to earn any revenue. The difficulties have been overcome to some extent by making it possible for customers to make use of the local loop for access to the trunk networks owned and operated by other carriers, where the traffic may be carried at a lower charge.

In the simplest case, a customer wishing to make a long-distance call will decide to use Cable and Wireless. From a telephone connected to the BT network, first of all, he will dial a Cable and Wireless access code, followed by his own unique code (PIN) to enable the call charge to be billed to the right customer, and then the destination telephone number. This is a generous sequence of digits to be dialled, to say the least, and so telephones have appeared with a programmable button that does nothing more than access a memory location in which may be stored the access code and the customer's PIN. The dialling sequence for a long-distance call over a Cable and Wireless line is now the same as for a BT call, with the addition of just one button press. However, the customer still needs to decide if the call is going to be cheaper via BT or Cable and Wireless, and this will depend upon his geographical location and the destination of the call. A BT local call is usually cheaper, but calls around the outskirts of the local call area may require a reference document to determine the least-cost routing (LCR) for any particular call.

Special adaptors, when inserted between the telephone and the BT socket and suitably programmed, will not only insert the access code and customer PIN, but also decide by which carrier the call will be cheaper, and route the call accordingly. This approach is the most obvious if a number of people will be making calls. Many private exchanges (PABXs) now offer automatic LCR built in to the exchange, thus obviating the need for an additional adaptor unit.

More recently, service providers have made use of the calling line identification (CLI) facility that transmits the caller's telephone number when a call is set up. This number uniquely identifies the user to the service provider, and the use of a PIN becomes unnecessary.

Integrated services digital network (ISDN)

While the 4 kHz voice channel is perfectly adequate for ordinary telephone conversations, it is quite limiting when the connection is used instead for data communication purposes. An increase in bandwidth will allow higher frequencies, and therefore higher data rates, to be accommodated. In order to meet the demands of the current age, where large amounts of data are being transferred between business users, and many people are requiring high speed access to the Internet, another network, running in parallel to the analogue PSTN, has been introduced. By using digital techniques, customers may benefit from the use of wider bandwidths, and therefore higher data-transfer rates. A different interface from the simple local loop just described is required, but customers are still able to initiate and receive calls via the PSTN. The advantages become apparent when an ISDN customer wishes to communicate with another ISDN customer, where the wide bandwidth available allows, for example, high-speed data transmission, high-quality speech, and video conferencing.

An ISDN connection is divided into channels of two different types. The first, the 'B' channel, carries digital data at a rate of 64 000 bits per second (64 kbit/s), where a bit is a Binary digIT. The second, known as the 'D' channel, is restricted to a data rate of 16 kbit/s, and is normally used for signalling purposes. Because the data channels are separate from the signalling channels, the data stream is uninterrupted by control signals, therefore allowing the highest possible rate of data flow.

There are two types of ISDN access, known as basic rate access, abbreviated to BRA, and primary rate access, or PRA.

Basic-rate ISDN provides a customer with two 'B' channels and one 'D' channel. With appropriate terminal equipment, this may use the existing ordinary telephone line to a customer's premises. The network operator may also find it to be more cost-effective to provide BRA to a customer who requires an additional telephone line than to install a new line.

Primary-rate ISDN provides for 30 'B' channels and two 'D' channels. This requires an upgraded line between the customer and the local exchange, since any existing telephone lines will be unable to operate at the required 2 million bits per second (2 Mbit/s) data rate.

Advantages of using ISDN

The wider bandwidth and higher speed of ISDN mean that applications that were once unthinkable have now become a reality. For example, many business meetings do not need the physical presence of the members, but could be effectively accomplished if all the parties were connected together by voice and by video. Two or more ISDN customers can do this by a system known as 'video conferencing'. The benefits are obvious – thousands of pounds may be saved in travel and accommodation costs and travel time.

Recent case studies have identified other areas where substantial cost and time savings can be made through the use of ISDN.

- A customer with branch offices in three other countries used to exchange computer files by sending floppy discs by mail or by courier. Now, by transmitting the files by ISDN, the data are up to date and readily accessible to all.
- A graphic-design company relied on mail, couriers, or modems for communicating their designs to customers. By switching to ISDN, they have been able to save substantially on courier fees, and have expanded their business by targeting potential customers who also have ISDN capability.
- The head office of a chain store has installed ISDN in all of its branches. This enables daily sales figures to be routinely retrieved from each branch by the head office and stock to be automatically replenished. It has also speeded up credit-card transactions due to the higher connection speed of ISDN.
- Many people whose work is information-based are now able to work at home because they are able, thanks to ISDN, to download computer files from their place of employment as quickly as if they were using a computer in the office. Apart from the time and money being saved due to reduced travelling (with environmental benefits as a spin-off), the employer can reduce the amount of accommodation required for his desk-bound staff.

Basic rate access

The original version of BRA, dating back to the mid-1980s, was known as ISDN 2. Subsequent harmonization with European standards has brought about a modified, and now current, version – ISDN 2e. The service is aimed at the medium-sized business market and arrives at the customer's premises via an ordinary telephone cable pair. At what is known as the 'U' reference point, this pair is terminated at a network terminating unit, or 'NT', which converts the two-wire network connections to a four-wire interface, where the transmit and receive paths have been separated. The customer may now connect any ISDN-compatible equipment to this inter-face, known as the 'S-bus'. If non-ISDN-compatible equipment, such as an ordinary telephone or modem, is to be connected, then a terminal adaptor ('TA') is connected to this point. Fig. 1.11 shows the general scheme. Under normal circumstances, power for the NT and for ISDN-compatible terminal equipment ('TE1') will be supplied by a power adaptor connected to the local mains supply.

Up to eight items of TE1 may be connected, in parallel, to the S-bus, and each one may be separately addressed by the network connection. This means that up to eight telephones, say, may be connected. The network

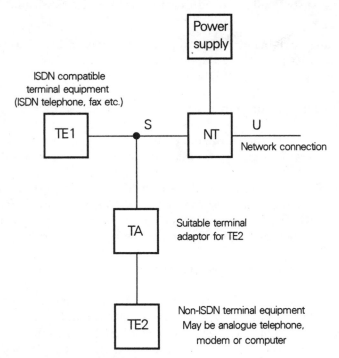

Figure 1.11   *ISDN 2e implementation at customer's premises.*

operator will allocate a block of consecutive telephone numbers to the connection, the last digit(s) of which will not be used for routing the call through the public network, but will be broadcast to all of the terminals connected to the S-bus. Each terminal will be programmed to respond to a specific number. The scheme is known as multiple subscriber numbering, or MSN.

In the case of a power failure, only one item of TE will be operative (normally a telephone for emergency use), the power being derived from the network.

While this level of complexity is acceptable to the medium-sized business community, it is rather a daunting prospect for the small business and for the domestic customer. This is a considerable market segment that nevertheless would embrace the idea of high-speed Internet access with open arms. Mindful of this market, BT has introduced more user-friendly forms of ISDN 2e, known as Business Highway and Home Highway.

In both cases, the network terminator and two terminal adaptors have been combined into one unit. This scheme is shown in Fig. 1.12. In this implementation, the customer has two familiar analogue telephone sockets to which may be connected ordinary telephones, fax machines and the like. There are also two digital connection points, primarily intended for one or two computers to be connected. Suitable terminal adaptors will be required for the digital connections, which could be either internal computer expansion cards, or separate external units. Each analogue socket has a separate telephone number, and a third number is provided for the direct digital access.

Not all of the sockets may be used at once, however. The permissible combinations are shown in Fig. 1.13. Each connection point, or 'port', uses 64 kbit/s of the available 128 kbit/s bandwidth. However, to achieve the highest possible data rate, the entire bandwidth may be devoted to one computer link, in which case, no other connection may be used simultaneously. It should be noted that in order to convert to a high-speed Internet connection, the user must contact his ISP (Internet service provider) who will provide a different telephone number for ISDN access.

Unlike ISDN 2e, BT Highway does not provide a power source for ISDN-compatible equipment. Any devices connected to the digital ports will therefore need their own power supplies.

BT Home Highway differs from both ISDN 2e and Business Highway in that multiple subscriber numbering – MSN – is not supported. However, the requirement for this feature is unlikely in domestic circumstances.

Primary rate access – ISDN 30

ISDN 30 is aimed at the large business customer, with more than ten analogue lines. A single ISDN 30 connection, usually made by fibre-optic link

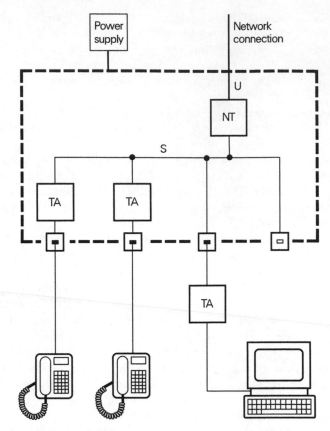

Figure 1.12   *BT Highway implementation at customer's premises.*

between the customer's premises and the local exchange, will support up to 30 ISDN channels of 64 kbit/s each, although the minimum configuration is eight such channels. At the customer's premises, these channels would normally be terminated in a suitable telephone exchange that would be able to handle the complex procedures involved. Because the precise method of installation will be largely manufacturer-dependent, as will the facilities available and the methods of programming, installers would need to equip themselves with the specific knowledge required by attending a manufacturer's course. For those readers wishing to learn more about ISDN, Appendix 2 lists some of the titles available.

Analogue telephone 1	Analogue telephone 2	Digital 1	Digital 2
✓	✓	✗	✗
✓	✗	64 k	✗
✗	✗	64 k	64 k
✗	✗	128 k	✗

Figure 1.13   *BT Highway permissible access combinations.*

'700 Series' U.K. 1959

2
Connectors and cables

Plugs and sockets – the BT 'New Plan' connection system

Most people will by now be familiar with the plug-and-socket system introduced by BT. The connection may be either six-way, arranged 'in line', or four-way, in which case, the two outside contacts are missing. For practically all domestic installations, and most business applications, four-way connectors are quite adequate. There is a large variety of different sockets ('line jack units', LJUs) available, bearing different code numbers, which begs some explanation.

The BT master socket

LJU1/1A, LJU2/1A, LJU3/1A

When the modern system of telephone wiring was introduced, it was intended that pieces of telephone apparatus could be plugged into, and unplugged from, the telephone line, in much the same way as electrical appliances are connected and disconnected to and from the electricity supply. It also means that a subscriber may choose not to have any instruments connected to the line – he may 'opt out of service'.

Previously, telephone instruments were hard-wired to the network, and the integrity of the line could be tested from the exchange by measuring the impedance (= AC resistance) of the local loop; that is to say, the combined impedance of the line, the ringer and the associated 'bell capacitor', all in series. A broken line would appear as an infinite impedance, a shorted line would produce a very low impedance, while a good line would exhibit an impedance in the order of 2000 or 3000 Ω (2 or 3 kΩ). In order to maintain this testing facility, the bell capacitor has been relocated

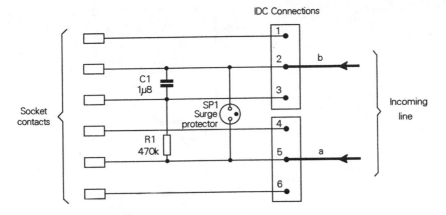

Figure 2.1   *Master socket – circuit diagram.*

in the master socket, and an additional component, a 470 kΩ resistor, is connected across what now becomes the bell terminals (see Fig. 2.1). Even if no telephone is connected to the line, there is still an AC circuit of about 470 kΩ (the line and capacitor resistance being relatively small), and the line testing facility is maintained. The resistor is known to BT as the 'opt out of service' resistor.

The main reason, however, for locating the bell capacitor in the master socket, thus allowing the bell signal to be distributed separately, is to permit telephones to be connected in parallel. If this were done on a line terminated without the master socket, all of the telephones would 'tinkle' sympathetically when one telephone was being used to dial out, because the bells would 'see' the fluctuations on the line due to the dialling pulses as being similar to a ringing signal. The master socket's internal bell capacitor allows a third 'anti-tinkle' connection to be made to each telephone. When one telephone is being used to dial out, it effectively shorts out the bell circuit by connecting a low-value resistor across the bells, rendering them all temporarily inoperative.

One additional component, the surge protector, is also included. This consists of a spark gap inside a gas-filled glass envelope. The device presents an infinite impedance and resistance unless the voltage across its terminals rises above a critical level (usually between 200 and 260 V), at which point, it conducts heavily and prevents the voltage from rising further. Note that a surge protector across the line does not provide lightning protection – it only prevents the voltage between the two wires of the line rising excessively. In the case of a lightning strike, both wires of the line would almost certainly receive the impulse together and rise above earth potential. This may also be damaging, but effective protection requires two surge protectors, one from each line to earth.

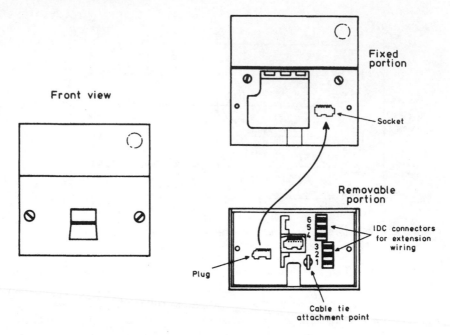

Front view

Fixed portion

Socket

Removable portion

IDC connectors for extension wiring

Plug

Cable tie attachment point

Figure 2.2   *NTE5 general arrangement.*

Network terminating equipment 5A/B (see Fig. 2.2)

A later development of the '–/1A' type of master socket, the NTE 5A/B (BT 'Linebox') was introduced to overcome the difficulty of connecting installation cable via the plug and socket arrangement designed for the flexible lead and special plug of telephone apparatus. Even when accomplished with special adaptors, the result is clumsy and obtrusive.

The front plate of the NTE 5A/B is split into two parts, the lower half of which may be removed by the subscriber. Extension wires may be attached to the terminals on the back of the removable part, which is then plugged back into the fixed part of the unit. The wiring may be therefore completely concealed in cavity walls or ducting, and the whole effect is much neater. Two versions are available: the 5A, which has a standard equipment socket on the outside of the removable part, and the 5B, which does not. The NTE 5A/B also has provision for a choice of surge protector – the single protector (BT Type 11B) across the incoming line for cable feeds, which are underground and are not likely to suffer lightning strikes, and the twin protector (BT Type 21A), which requires an additional earth connection to the unit (see Fig. 2.3). In a later version of the linebox, the earth connection has been omitted so that it is only possible to fit a type 11B surge protector; and only connections 2, 3, 4 and 5 appear at the IDC terminations.

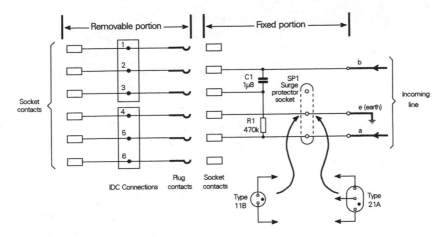

Figure 2.3 *NTE5 circuit diagram.*

Note: The installation and connection (other than by means of the telephone socket or as described above for the subscriber's connection to the linebox) is the responsibility of the PTO (public telecommunications operator). Once installed, the master socket or equivalent may not be moved or altered except by an agent of the PTO. If it is wished to install extension wiring, there must already be a master socket installed on the premises. If one has not been installed, the PTO will do so upon request and will charge for it.

If a user has a 'party line' – 'shared service' (a line shared with another user), the 'New Plan' system cannot be installed. However, the PTO is under an obligation to provide master sockets upon request, and if a master socket is requested, the PTO has no option but to convert the line to a single user line.

The secondary socket – '–/3A'

Once the master socket has been installed by the PTO, there is no limit, effectively, to the number of secondary sockets that may be installed. (There is, however, a limit to the number of pieces of telephone apparatus, and to the additional line length that may be connected – see Chapter 3.)

The secondary socket contains no additional components as in the case of the master socket.

The PABX master socket – '–/2A'

There is a third configuration, which is normally only used in some business systems where a PABX (private automatic branch exchange) has been installed. In this case, each extension number has its own 'local loop', and

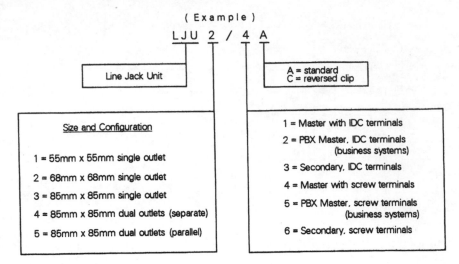

Figure 2.4   *LJU designations.*

therefore, a 'master socket' is required. However, since neither protection against high-voltage impulses nor line testing is needed, the only component provided is the bell capacitor.

Socket styles, sizes, and configurations

Figure 2.4 shows how the designations of the various versions of line jack units are made up.

After the common prefix 'LJU' comes a number that denotes the size and configuration. 1, 2 and 3 are all single outlets with 1 being the smallest (55 mm × 55 mm). It is also unique in that the PCB (printed circuit board), components and connector are mounted on the fixed portion, the removable part being simply a faceplate.

Size 2 is the size most commonly used for surface mounted applications, while size 3, and dual outlets size 4 and 5 are of the same dimensions as 240 V AC electrical 13 A sockets, switches, etc. This type is more suited to flush-mounted applications.

A further numerical digit follows the '/' stroke, and this digit (1–6 inclusive) denotes master or secondary, as well as the type of terminal for wiring connections. IDC (insulation displacement connectors) are most commonly used as they are less expensive and considerably faster to install. However, if only a small installation is contemplated and the special tool is not available, screw terminals are the obvious choice. This choice may also be indicated if non-standard wire sizes are to be used

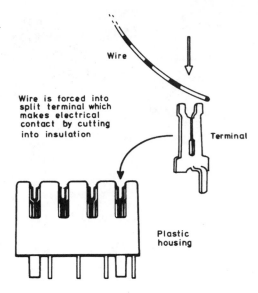

Wire

Wire is forced into
split terminal which
makes electrical
contact by cutting
into insulation

Terminal

Plastic
housing

Figure 2.5 *Principle of the IDC joint.*

since the integrity of the IDC joint is dependent upon the correct wire size being used. The principle of the IDC joint is shown in Fig. 2.5.

The last character of the designation is a letter. 'A' is used for the standard product, while the letter 'C' indicates that the plastic retaining clip of the mating plug is catered for on the opposing side of the connector to normal. In business applications, this may be used to ensure that only certain types of telephone may be plugged into particular sockets. (PABX systems commonly cater for both standard telephones and proprietary 'key-stations' with features that require special wiring. The use of non-compatible sockets for standard phones and keystations prevents confusion.)

In addition to the LJUs and the NTE 5A/B, there is also a compact, surface-mounted secondary socket available for use where space is limited. Note that this side-entry socket does not permit the use of two-way adaptors.

Various other forms of sockets also exist, some of obsolete design (LJU11; 621A (master), 622A (secondary)), and some with special features such as the enclosed line jack where the plug and socket are inside the box to discourage removal.

Occasionally, a need will be found for a block terminal 81A, shown in Fig. 2.6. It consists of a PCB on which are mounted all the components of a master socket except for the socket itself, and is known as a 'mastering unit'. Because of its compact dimensions, it can be fitted inside trunking

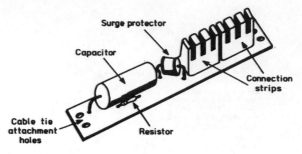

Figure 2.6   *Block terminal 81A.*

or other housings using the adhesive foam strip supplied, and can be useful when fitting bells to two-wire lines.

There are also modified sockets acting as 'privacy adaptors', or offering call-barring facilities. These are considered in Chapter 8.

'Break-in' adaptors – LJU10/7A and LJU10/11A

Commonly used with fax machines, these adaptors allow a telephone to be plugged into the same outlet and used to make and receive voice calls in the normal way. However, the fax machine or other device can disconnect the telephone outlet by operating relay contacts and thus take control of the line. This prevents interference with the data signal from extraneous noise picked up by the telephone handset.

The two types are outwardly identical, apart from the type numbers, and consist of a small box with an integral line plug and socket outlet, with a flying lead attached, with spade terminals. The connection scheme for each type is shown in Fig. 2.7. The housings are not very robust, and replacement of the adaptor is often necessary because of physical damage. Care must be taken to ensure that the correct replacement is fitted.

Mating plugs for line jack units

Description

A series of IDC plugs has been developed for use with the LJU and NTE5 types of socket. Both four- and six-way versions are available, and the latch may be on the left or right side. Table 2.1 shows the coding of the various types, which are depicted in Figure 2.8.

The system was developed by BT to fulfil a number of requirements. One such requirement was that the system should be useable both for line connections and for the connection of the handset cord to the telephone. In order to prevent misconnection, the latch is on one side of the plug for the handset version and on the opposite side for the line connector.

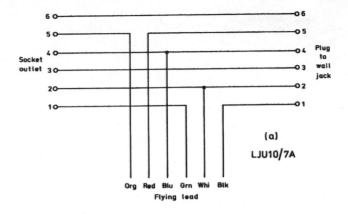

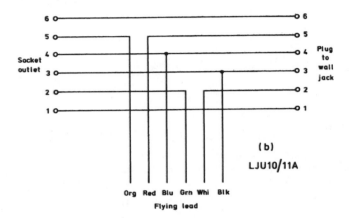

Figure 2.7 *Connection schemes – LJU10/7A and LJU10/11A.*

Table 2.1 *Coding of line and handset plugs*

Code number	Description
430A	Four-way handset plug (latch on left)
431A	Four-way line plug (latch on right)
630A	Six-way handset plug (latch on left)
631A	Six-way line plug (latch on right)

Since this original conception, handset cords are now usually equipped with a smaller series of connectors, described later. The handset type of plug may thus be used for other purposes, such as for the connection of a 'key telephone' as previously described.

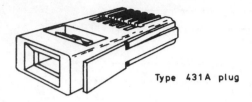

Type 431A plug

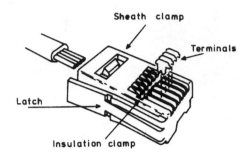

Figure 2.8   *Plugs for LJUs.*

Assembly

It must be stressed that this series of plugs is designed for use with CW1311 'D'-profile flexible cordage, and for no other.

While there may be occasions where it is desired to terminate other types of cable with these plugs, it is strongly recommended that it is not attempted. Even if an attempt is apparently successful, reliability is in serious doubt.

There are various types of specialized crimping tools designed for use with these connectors. Being normally of plastic construction, they are relatively inexpensive, and their use almost guarantees a high success rate in cable assembly.

There are some DIY packs of plugs, together with an 'assembly tool', which is a piece of cast metal. Used in conjunction with a pair of parallel-jawed pliers, it is possible to terminate a few cables successfully but with considerably more difficulty than with the tool described above.

Whichever method is chosen, the preparation remains the same. The outer sheath of the cable to be terminated is removed to expose 15 mm of the inner conductors, still with their own insulation intact. The conductors should be carefully separated by a slight amount to enable them to slide easily into their own channels in the connector. The prepared end is then gently inserted into the connector, pushing it fully home, ensuring that it is the right way round. The rounded surface of the cordage should be on the same side of the connector as the visible metal contacts.

The contacts and cable retaining clips are then crimped in one operation. Once the connector has been crimped, it is not possible to inspect the joint visually, and so it is doubly important that these guidelines and the instructions accompanying the crimp tool be followed carefully.

FCC68 specification connectors – 'Western Electric' RJ11 and RJ45

A smaller series of connectors, used in the United States for wall-socket connections, and for computer networking purposes, will also be encountered. These are commonly used for handset connection, for connecting a line cord to some apparatus, and occasionally for other purposes. They are made of clear plastic, and the retaining latch is centrally located. In some cases, the latch lever will have been shortened, and the plug can only be removed from the socket by first disengaging the latch with an instrument such as a jeweller's screwdriver.

The smallest plug has only four contact positions and is only supplied with all four contacts loaded. A slightly wider plug has six contact positions, available with either four or six contacts loaded. In the latter case (6/6), it is designated RJ11. A third style, which is wider again, has eight contacts, and is designated RJ45. This type is commonly used in structured cabling systems (see Chapter 6). The largest plug has ten contact positions, and is unlikely to be encountered in telephone applications.

For assembly purposes, separate crimp tools are available for each size, and some combination sets are also marketed.

Block terminals and distribution boxes

Block terminals are used for joining cables together, for splitting multi-core cables into two or more smaller cables, and for forming 'T' junctions.

Distribution boxes are also used for joining cables, but in this case, a multipair cable (for example, from an exchange) is made off on one set of terminals, the outgoing cables (to the extensions) are made off on another set of terminals, and the connections between the two are made by means of jumper links. In this way, cable terminations are considered to be permanent, and any modifications that need to be made are carried out by replacing the jumpers. The difference between the two concepts is illustrated diagrammatically in Fig. 2.9.

Block terminals

Block terminals, until fairly recently, have been of the screw terminal type, where the wire ends to be joined are first stripped, twisted together, then formed into a loop, and then trapped beneath the head of a screw. This method is fiddly and time-consuming, and does not lend itself readily to

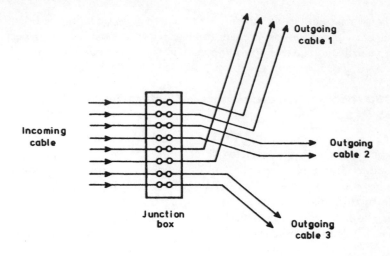

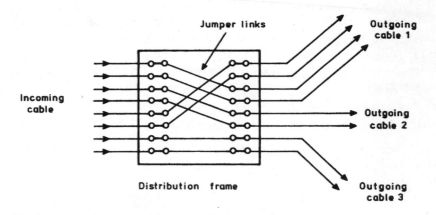

Figure 2.9 *Junction box – distribution frame comparison.*

modifications later. However, many thousands of such chocolate-coloured boxes with their distinctive shape (basically rectangular, but with slightly curved sides and ends, and rounded corners) are still in use today. They are available in a number of different sizes, up to about 20 pairs. This method has now largely been superseded by the IDC (insulation displacement connector) system, sold in this country by Krone (UK) Technique, and in extensive use.

The advantages of this type of connection have already been mentioned. Various wire sizes can be accommodated, but note that if two wires are

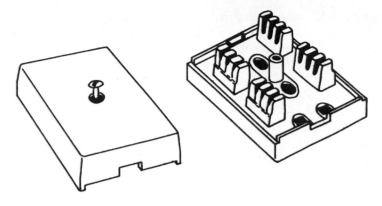

Figure 2.10 *Block terminal 77A.*

inserted into the same terminal, they must be of the same diameter. For 22 Standard Wire Gauge (SWG) and greater, only one wire may be inserted.

It would not be appropriate to describe all of the available types of junction box, but here is a review of the more common and useful types.

Block terminal 77A consists of two parallel rows of six IDC contacts per row, mounted on a PCB. The contacts of one row, numbered 1–6, are connected by means of the tracks of the PCB to the second row contacts numbered 1–6, respectively. This arrangement allows up to four three-pair cables to be parallel-connected. See Fig. 2.10.

Block terminal 77B is similar, but connections are made by screw terminal blocks instead of IDC connectors. This allows a wider range of wire sizes to be accommodated and also means that it can be installed without the use of special tools.

Block terminal 78A is similar to the 77A, but provides for eight wires (four pairs) instead of six (three pairs).

Block terminal 80A is a useful three-wire interface, where connections have to be made from non-standard diameter cables, particularly 'dropwire', to internal cabling.

Block terminal 84A is electrically the same as the 77A, but has IDC on one side and screw terminal blocks on the other. Like the 80A, it is particularly useful when forming an interface between standard and non-standard wire sizes, but catering for six wires rather than three.

All of these block terminals, and a few other configurations, are housed in a small rectangular box measuring 56 mm × 42 mm × 23 mm. They are suitable only for internal use.

Block terminal 66B measures 125 × 90 × 42 mm, and is intended for external use. It contains two five-way screw terminal blocks similar to the familiar 'chocolate block', and is intended for terminating and joining external cable such as 'dropwire'. The ends of the terminal screws that are in contact with the wire are pointed. It is intended that the wires should not be stripped prior to insertion, as is normally the case. Rather, the screw penetrates the insulation to make the contact. For this reason, only one wire may be placed beneath each screw. If a T-connection is desired in an exposed location, it is recommended that insulation-displacement jelly-filled splices be used, housed within a 66B enclosure.

Splices are in the form of small, clear plastic, jelly-filled two-part capsules. They are available in two- and three-way versions. In use, wires are inserted into the access holes (one wire per hole), and the two parts of the capsule are then squeezed firmly together using a large pair of pliers. These splices are useful for a number of purposes:

- when extending wires within connecting boxes of one sort or another when a short extension is needed;
- as a neater alternative to a 77A block terminal when only a few wires are involved and there is a convenient socket or other housing nearby;
- as a weatherproof joint.

Distribution boxes

Modern IDC distribution boxes are of modular construction. The housing consists of a base and a removable lid with sides. The sides of the lids are equipped with 'knock-outs', or slide-out panels, to allow wiring to pass through. Several boxes of the same size can be mounted side by side, with adjacent knock-outs removed to give greater capacity. Some of the most useful configurations are listed below.

IDC terminal strips consist of two parallel rows of terminals in the same moulded plastic block. The terminals in one row are internally connected to the corresponding terminals in the second row. If this connection is fixed and permanent, the strip is referred to as a 'connection strip'. If, however, the connection is made through contacts that may be broken by the insertion of a monitoring plug, or special disconnection plug, the strip is called a 'disconnection strip'. The commonest available types have a capacity of 10 pairs – that is, each of the two parallel rows each has 20 contacts. The capacity of each terminal is as listed in Table 2.2.

Box connection 201 measures $170 \times 120 \times 35$ mm. It contains two separate low-profile, 10-pair, IDC connection strips type 241B and a common earthing strip. These boxes are found to be most useful when terminating or joining a multipair cable of up to 20 pairs. For example, a 10-pair cable carrying up to five extensions might be terminated on one strip, and a number of smaller cables terminated on the other strip. The two strips are linked together by means of jumper wires. The internal layout is shown in Fig. 2.11. The box connection 201/10 is identical but is supplied with only one 241B connection strip.

Box connection 222 measures $170 \times 120 \times 65$ mm. It provides the ability

Table 2.2 *Wire sizes in IDC terminals*

Diameter (mm)	Nearest SWG	Number of wires per contact
0.4	27	2
0.5	25	2
0.6	23	2
0.65	23	2
0.7	22	1
0.8	21	1
0.9	20	1

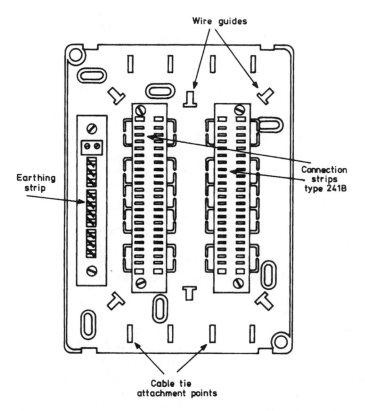

Figure 2.11 *Connection box 201C internal arrangement.*

to terminate up to six pairs with screw connections on one side and IDC on the other. The IDC strips are type 244A, which allow both for disconnection and for testing. Several commoned earthing points are also provided. This box is ideal for use with small systems providing termination facilities at the PSTN connection point.

Box connection 251 measures $210 \times 160 \times 90$ mm. It may contain up to five 10-pair connecting strips and is commonly used as the distribution point for the extension wiring of PABXs.

Box connection 301 measures $320 \times 210 \times 120$ mm. It may contain up to ten 10-pair connecting strips and is commonly used in the same way as a 251 box where a greater capacity is required.

Accessories for the above connection boxes include labelling strips, earth terminal points, wiring guides and magazines for surge protectors. Labelling strips are available in two sorts: one type that clips on top of a connecting strip and can be hinged up to expose the terminals, and a variety that is inserted into a slot normally occupied by a terminal strip. Both types should only be marked in pencil to make future changes easier.

The surge protector magazines (Type 5B) provide a convenient way of protecting against surges on external lines, either from the PSTN or from any extension wiring which goes outside the building, especially overhead. Note that the additional height required when a 5B is plugged into the top of a 10-pair strip means that they are unsuitable for use in low-profile boxes.

All of the above types are only suitable for internal use. Weatherproofed versions are available from specialist suppliers.

Cable types and colour codes

Internal cables

For internal wiring, 0.5 mm tinned copper wire covered with PVC insulation and an overall PVC outer sheath is normally used.

CW1293

Although not recommended for new installations, Type CW1293, identifiable by its cream outer sheath, has been extensively used and is still available. There are two types – 'quad' cable and 'paired' cable.

In the 'quad' cable, the four wires are coloured blue, orange, green, and brown. The blue and orange wires were normally treated as a pair (blue = a-wire, orange = b-wire), and the green and brown were used either as another pair, or for signalling or earthing purposes.

In the 'paired' cable, the wires are arranged in twisted pairs, the colours being in accordance with Table 2.3. Identification of the white, red, black etc. wires is dependent upon successfully determining which other colour is twisted with it. As the number of twists per unit length is not great, an appreciable amount of outer sheath must be carefully removed to be certain of correct identification.

Table 2.3 *Wire identification colours for CW1293*

Cable pair	Colour of insulation		Cable pair	Colour of insulation	
	a–wire	b–wire		a–wire	b–wire
1	WHITE	BLUE	21	WHITE–blue	BLUE
2	WHITE	ORANGE	22	WHITE–blue	ORANGE
3	WHITE	GREEN	23	WHITE–blue	GREEN
4	WHITE	BROWN	24	WHITE–blue	BROWN
5	WHITE	GREY	25	WHITE–blue	GREY
6	RED	BLUE	26	RED–blue	BLUE
7	RED	ORANGE	27	RED–blue	ORANGE
8	RED	GREEN	28	RED–blue	GREEN
9	RED	BROWN	29	RED–blue	BROWN
10	RED	GREY	30	RED–blue	GREY
11	BLACK	BLUE	31	BLUE–black	BLUE
12	BLACK	ORANGE	32	BLUE–black	ORANGE
13	BLACK	GREEN	33	BLUE–black	GREEN
14	BLACK	BROWN	34	BLUE–black	BROWN
15	BLACK	GREY	35	BLUE–black	GREY
16	YELLOW	BLUE	36	YELLOW–blue	BLUE
17	YELLOW	ORANGE	37	YELLOW–blue	ORANGE
18	YELLOW	GREEN	38	YELLOW–blue	GREEN
19	YELLOW	BROWN	39	YELLOW–blue	BROWN
20	YELLOW	GREY	40	YELLOW–blue	GREY

CW1308

To overcome the problem of wire identification, a new type of cable, CW1308, was introduced, which is recommended for use in new installations.

In the new cable, identified by its white or black outer sheath, every wire is marked throughout its length with two colour bands: a base colour and a tracer colour, in the ratio 3:1, repeated over a 25-mm length.

Each wire of the pair uses the same two colours, but the base and tracer colours are reversed to permit distinction between them. Using the colours blue, orange, green, brown, grey, with white, red, black, yellow, and violet, up to 25 pairs can be made up as shown in Table 2.4. Throughout this book, the base colour is identified in upper-case letters and the tracer colour in lower-case letters.

However, cables commonly used for distribution wiring use the first 20 pairs as a unit, repeated as necessary. In addition, to simplify the distribution of a functional earth, a 1.5-mm earth wire is included in most cables with 10 pairs or greater.

Table 2.4   *Wire identification colours for CW1308*

Cable pair	Colour of insulation a-wire	b-wire	Cable pair	Colour of insulation a-wire	b-wire
1	WHITE–blue	BLUE–white	16	YELLOW–blue	BLUE–yellow
2	WHITE–orange	ORANGE–white	17	YELLOW–orange	ORANGE–yellow
3	WHITE–green	GREEN–white	18	YELLOW–green	GREEN–yellow
4	WHITE–brown	BROWN–white	19	YELLOW–brown	BROWN–yellow
5	WHITE–grey	GREY–white	20	YELLOW–grey	GREY–yellow
6	RED–blue	BLUE–red	21	VIOLET–blue	BLUE–violet
7	RED–orange	ORANGE–red	22	VIOLET–orange	ORANGE–violet
8	RED–green	GREEN–red	23	VIOLET–green	GREEN–violet
9	RED–brown	BROWN–red	24	VIOLET–brown	BROWN–violet
10	RED–grey	GREY–red	25	VIOLET–grey	GREY–violet
11	BLACK–blue	BLUE–black			
12	BLACK–orange	ORANGE–black			
13	BLACK–green	GREEN–black			
14	BLACK–brown	BROWN–black			
15	BLACK–grey	GREY–black			

Cables to these specifications are available in 1-, 2-, 3-, 4-, 6-, 10-, 12-, 15-, 20-, 25-, 40-, 60-, 80-, 100-, 160- and 320-pair configurations. In practice, 3-, 4-, 6-, 10, 20 and 40-pair cables will suffice for most purposes, with 3-pair, 6-pair and 10-pair being the most useful.

CW1316

All the above cables have a circular cross-section. A flat six-core cable with the same wire identification colours as CW1308 is also available, designed for under-carpet use. This cable has a particularly tough outer sheath and carries the designation CW1316.

CW1311

For connecting items of telephone apparatus to the socket, four-way or six-way D-shaped cordage is used. The conductors are stranded (seven strands of 0.152 mm diameter) to ensure flexibility. The internal colours of the individual wires are shown in Table 2.5. There are two problems of which one should be aware when dealing with this type of cable. The first is that by no means does all D-profile cordage use these wire colours. This commonly applies to equipment of Far-Eastern origin. The second is that the wire colours will be reversed when the other end of the cable is examined.

Table 2.5 *Wire identification colours for CW1311*

Pin	Purpose	Colour	
(1)	*	(Black)	(six-way only)
2	b-wire (line)	White	
3	Bell anti-tinkle	Green	
4	Earth	Blue	
5	a-wire (line)	Red	
(6)	*	(Orange)	(six-way only)

* The outermost wires (pins 1 and 6) may be used for signalling, switching, or low-voltage power-supply distribution in business systems.

It is therefore unwise to make any assumptions concerning the designation of any particular wire colour without cross-checking.

CW1600

CW1600 is a specification for a limited fire-hazard cable with low smoke and fume emissions, and a fire-barrier tape. It also features a foil screen. The form of construction limits the minimum bending radius to eight times the diameter of the cable. The electrical characteristics and colour coding are as for CW1308.

CW1044

CW1044 is the specification of a single-stranded conductor with a cream outer sheath, marked with the words 'TELECOMMS FUNCTIONAL EARTH'. Various conductor sizes are available from 1.5 mm^2 upwards.

External cables

Cables intended for external use differ in construction because they need:

- to withstand higher levels of mechanical stress due to being buried or exposed to wind, or other forces;
- to have a higher degree of protection against moisture;
- to be able to withstand wider temperature ranges than cables for internal use.

CW1378 (Dropwire)

Cables that are intended for overhead installation without extra support ('dropwire'), are provided with additional insulated steel wires combined with the conductors. These wires are not intended for electrical use. The colour coding for the two-pair 'dropwire No. 10' is shown in Table 2.6.

Table 2.6   *Wire identification colours for Dropwire No. 10*

Wire colour	Pairing
White–orange	Pair 1
Black–green	Pair 2
3 × yellow or red	Steel strainers: not used electrically

CW1128

Cables for underground routing need to be protected against impact, abrasion, and even rodent damage. Beneath the tough outer sheath of such cables lies the armouring of galvanized-steel wires that completely envelope the remainder of the cable. The functional wires are separated from the armouring by a further sheath that is filled with petroleum jelly to exclude moisture.

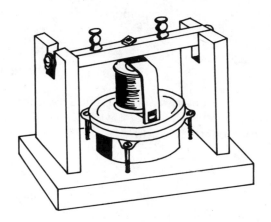

'Gallows' U.S.A. 1875

3
Installation techniques

Accommodation

The starting point for any installation work is to assess the proposed location of the equipment and cabling for suitability from three different aspects:

- environmental
- access
- cable distances and routing.

From the environmental point of view, electronic equipment obviously should not be installed in areas that may be exposed to the weather or suffer from dampness, excessive heat, dust, or corrosive fumes. In addition, there may be special conditions prevailing that require consideration such as the presence of flammable or explosive materials, or the accommodation may be shared with other services.

It should be noted that installations in hazardous areas, in the proximity of electricity generating plant or substations, or where it is necessary for cabling to pass through firewalls, are subject to special requirements and recommendations. If any of these conditions apply, then BS6701 should be consulted.

Access to the areas concerned should be considered, with particular regard to the safety of installation, operation, and maintenance personnel. For example, junction boxes should not be sited above or behind doors. In addition, convenience of access should be maximized, ensuring that distribution frames are at a convenient working height, and that illumination of the areas of proposed work is adequate.

The length and routing of cables need special consideration. There may be a limit on the length of cable between the central control unit (CCU) and the operator's console, if any. Similarly, there may be a limit on the cable length between the CCU and the extension sockets. Clearly, the site chosen for the CCU must take these factors into account, as well as the need to route PSTN lines and extension lines, and to provide electrical power to the system.

Cabling

General

A well-planned route for the necessary cabling will clearly pay dividends, but as might be expected, the conflicting requirements of neatness, protection from mechanical damage and avoidance of contact with electrical supply wiring, conduits, gas pipes, heat and dampness, together with accessibility for maintenance and economy in materials and labour will lead inevitably to some compromises being made. In general, cabling will follow natural lines of architectural features – door and window frames, corners of rooms, etc., and, for neatness, exposed cabling will normally be either horizontal or vertical.

There may be a temptation to re-use existing telephone wiring that is redundant or will become so as a result of the new installation. Such a course of action needs to be considered very carefully, as the existing wiring may be faulty, may not meet the statutory requirements in one or more ways, and may have unexpected branches or joints. It is far better to re-wire completely than to risk the consequences of such problems. Except where cables have been plastered into walls, they can normally be used as 'pull-throughs' for the new cables.

The planning of the cable routing will also need to take into account the positions of the sockets and any junction boxes or distribution points. These should be sited to provide the best possible access while being protected from mechanical damage, especially from furniture and cleaning equipment.

Proximity to other cables – mandatory requirements

BS6701 makes a distinction between telecommunications circuits, low-voltage (LV) electricity supply cables, and electricity supply cables.

Definitions: A telecommunications circuit is one that operates at a voltage not exceeding 150 V DC, or 100 V AC constant, 175 V AC peak. Low-voltage electricity supply cables are those that operate at a higher voltage than 50 V AC or 120 V DC to earth, but not exceeding 600 V AC or 900 V DC to earth. Electricity supply cables are those that operate above 600 V AC or 900 V DC to earth.

Proximity to other telecommunications circuits

Within the voltage limits defined above, telephone wiring may be installed in close proximity to other telecommunication circuits and may share the same duct, conduit, or trunking. The installer should be aware, however, of the possibility of interference between adjacent circuits, and an increase in the distance between the circuits may be necessary to overcome the problem. Furthermore, the isolation of all cabling must be appropriate for the highest of the operating voltages. It should be noted that the insulation of both CW1293 and CW1308 telephone cables is rated at 80 V AC.

Clearly, where other telecommunication cables are present, and their operating conditions are unknown, the safest course of action is to treat them as LV electricity supply cables.

Proximity to LV electricity supply cables

This category will include both single- and three-phase electricity supply cables normally found on consumers' premises. In these cases, there is a mandatory requirement to maintain a spacing of at least 50 mm. If this spacing cannot be maintained, for example at crossing points, a non-conducting divider should be placed between the telecommunication cable and the electricity supply cable.

However, the 50-mm spacing requirement does not apply if the LV cables are enclosed in a separate conduit or trunking, which, if made of metal, must be earthed, or if the LV cables are of the earthed armoured or mineral insulated type.

It is expressly forbidden for telecommunication cables and any electricity supply cables (LV or otherwise) to share the same conduit or duct.

Proximity to electricity-supply cables

For the higher-voltage electricity-supply cables, a minimum spacing of 150 mm applies, although this distance may be reduced to 50 mm, provided that a rigid non-flammable insulator is in place.

Internal cables

There are several methods of attaching cables to walls and other surfaces. If the surface is suitable, possibly the quickest, simplest and neatest is to use a staple gun (see Chapter 9) loaded with staples of the correct size and colour – white staples are available to blend with white CW1308. The body of the staple gun may be used as a gauge to maintain an even distance between attachment points. To avoid kinks in the cable, 'wipe' the cable first to straighten bends before fixing.

The main danger with the use of the staple gun is stapling through the cable. This accident may short two or more conductors, or even sever

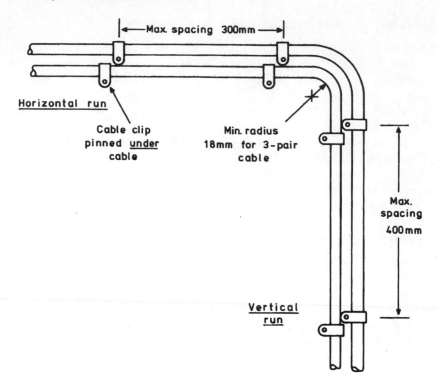

Figure 3.1   *Surface attachment of cables.*

them, and this is to be avoided! The cable should rest snugly in the guide of the staple gun before releasing the staple.

On masonry or other hard surfaces, and in places where there is insufficient space to use a staple gun, plastic cable clips will be found to be the most satisfactory – white for internal cables and black for external.

The recommended minimum spacing for staples or cable clips is 300 mm on an otherwise unsupported horizontal run, 400 mm on a vertical run. On a supported horizontal run – such as across a ledge – the spacing may be extended to 1 metre (m), but it is difficult to keep the cable tight with this spacing. These recommended spacings should be reduced on uneven surfaces.

The minimum bend radius recommended is four times the diameter of the cable, which, for the CW1308 three-pair, is 18 mm. Fig. 3.1 shows the preferred method of attachment at bends.

Where wiring cannot be readily concealed, the use of PVC trunking usually provides the neatest solution, especially where several cables are routed side by side.

External cables

The type of external cable should be selected according to the cable routing
– CW1378 for overhead and surface mount use, CW1128 petroleum jelly
(PJ) filled for underground ducts, or PJ armoured cable where the cable is
underground and not protected by a duct.

Overhead cables

It may be necessary to install an overhead cable between two buildings.
Copper cables do not have sufficient tensile strength for this purpose, and
so additional support is required. CW1378 dropwire includes additional
steel 'wires' within the cable structure, which provides this support. These
additional wires are not terminated, either electrically or mechanically.
Instead, the cable is mechanically attached at the ends of each span by
means of a 'dropwire clamp', which is a loop of steel wire with a long,
helically-wound tail (see Figs 3.2 and 3.3).

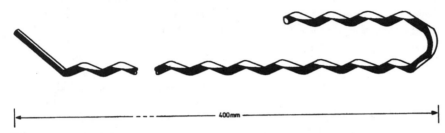

Figure 3.2 *Dropwire clamp.*

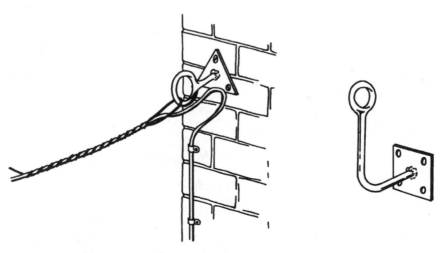

Figure 3.3 *Dropwire attachments.*

The tail is wrapped around the outside of the cable, where it provides a tight grip, the loop is passed through the fixing ring, and then the smaller end is wrapped over the clamp. Since the clamp can be attached at any point along a cable, not necessarily at the ends, two clamps may be used at intermediate points without the need for a cable joint. A small loop of spare cable is left between the two clamps to ensure free movement.

There is a choice of bracket to suit a variety of locations, including a right-angled version that is designed to be attached to a wall but provide clearance from overhanging eaves or gutters.

A steel catenary wire is moulded into the outer sheaths of heavier cables, in the range 10-pair to 50-pair. With this type of cable, only the ends and any intermediate junction points are mechanically attached by means of the catenary wire. Other attachment points are supported by a block-type bracket with a rubber clamping bush.

If the overhead telecommunications cable is to be supported by a separate wire, the quickest and neatest form of attachment is the cable tie. However, the material of most cable ties is subject to quite rapid deterioration on exposure to ultra-violet radiation, causing them to disintegrate. For areas exposed to sunlight, therefore, 'weather resistant' cable ties, which offer resistance to UV radiation, should be used.

Span lengths and heights (see Fig. 3.4)

The maximum recommended span length for overhead cables is 70 m, with a maximum sag of 735 mm that will be proportionately reduced for shorter spans. If the span crosses an area accessible to vehicular traffic, a minimum ground clearance of 5.5 m should be maintained. For other cases, a minimum clearance of 3.0 m is recommended.

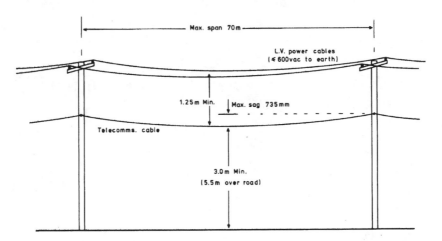

Figure 3.4   *Overhead cables – critical distances.*

Proximity to overhead electricity supply cables

On users' premises, it is unlikely that telecommunications lines will need to pass in proximity to cables operating in excess of 600 V AC. If there are, however, such cables in the vicinity, the situation needs to be assessed. As a rough guide, if the horizontal distance between the proposed telecommunications cable and the power line will be less than twice the height of the higher of the two lines, then special conditions may apply, and BS6701 should be consulted.

For LV power lines of less than 600 V AC to ground, which will be the more common situation, the requirements are rather more straightforward. Firstly, telecommunications lines must never pass over LV power lines, only beneath them, unless the telecommunications cable is insulated for that purpose. This is because power lines, being of more substantial construction, are considered to be less likely to break than telecommunications lines, and the likelihood of them ever coming into mutual contact is therefore reduced. Secondly, the minimum spacing between the two lines must be greater than 1.25 m from any unearthed conductor, or 1 m from any other part of the electrical equipment.

If there is a conflict between the minimum clearance height requirement, and the need for maintaining the minimum spacing to power lines, then either an alternative aerial route must be used, or the cable must be laid underground.

It should be noted that if it is proposed to make use of existing poles, it is essential that the agreement of the owner of the poles be sought.

Joints and T-connections in overhead cables

Overhead cables should never be joined in mid-span. The correct method is to join at an intermediate suspension point, such as a pole or corner of a building, using a connection box intended for external use. For dropwire, a Block Terminal 66B is a suitable weatherproof unit, and the individual wires of the cable may be joined using either the integral screw terminal connection blocks or jelly-filled IDC splices. The cables enter the box from the underside, and compression type saddle-clamps prevent the cables from being pulled out. For heavier multi-pair cable, and for totally waterproof underground use, various sizes of jointing sleeve are available. These are dome-shaped plastic enclosures, where the cable entry points may be sealed with a two-part resin compound, and are normally obtained as a kit of parts containing all necessary components and sealing materials. The wires are normally joined using jelly-filled splices.

Underground cables

The use of underground cabling, in spite of potentially higher installation costs, may be preferred for several reasons. There may be a need to avoid

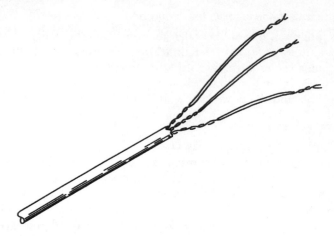

Figure 3.5   *Cable pairs identified prior to termination.*

power cables, there may be an existing duct for telecommunications cables, or there may be objection to the visual impact of overhead lines.

Cables for underground use will be petroleum jelly (PJ)-filled and may, or may not, be armoured.

Cables and ducts should be laid so that a minimum depth of cover of 350 mm is maintained. This depth should be increased to 450 mm in culti-vated ground and to 600 mm where the cable passes under a road. Where a duct is used, it should be sealed against moisture and gases at the point of entry into a building.

PJ-filled cables normally use the colour coding applicable to CW1293 paired cables, and thus an appreciable amount of outer sheath needs to be removed to ensure that the pairs are correctly identified. Carefully remove at least 500 mm of the outer sheath, gently extract each pair (blue/white, orange/white, green/white, etc.), and twist each pair together five or six times at the point where they emerge from the sheath. It may be helpful to twist the open ends together as well (see Fig. 3.5). Once the identification of the pairs is complete, the excess jelly may be removed. Take care that the insulation of the individual wires is not also removed – some cable types are rather prone to this kind of damage.

If joints or T-connections are required, a jointing chamber will need to be installed. These are fibreglass enclosures with metal covers, and serve to house jointing sleeves or other suitable forms of cable joint.

Building entry points

Door and window frames, and gaps beneath the eaves of buildings allowing access to the loft space are commonly used as entry points for cables. If a hole is drilled for the entry, ensure that it slopes downwards towards the outside so that moisture does not run in. For the same reason, cables entering a building should always do so by approaching the entry point from below. For cables originating from a higher point, it will be necessary to form a 'drip loop' to ensure that this condition is met. After the cable has been installed, the gap around it should be sealed with mastic sealant or other suitable compound.

For aesthetic reasons, it is normally desirable to terminate the external cable as soon as possible, continuing the wiring with internal-grade cable. The size of the cable will dictate the most appropriate connection box, but for a dropwire termination, a Block Terminal 77A, 80A, or 84A will be found adequate. It is inadvisable to mount any connection housing to a window sill, where pools of condensation may enter and cause corrosion.

Where overhead lines are in use, they are vulnerable to lightning strikes and other sources of high voltage. Where cables are terminated in a distribution frame, a surge protector may be conveniently installed to provide protection using a 5B magazine and 14A surge protector. Otherwise, a type 18 (wire-ended) surge protector may be installed in a Block Terminal 85A, remembering that an earth connection will be required at this point.

If surge protectors are installed in a location remote from the equipment that they are protecting, special consideration should be given to the method of wiring. If the earth connections are made in the manner shown in Fig. 3.6a, there is a danger that an instantaneous high voltage may be developed across c – e due to the inductance of the lead. This voltage would be added to the surge protector breakdown voltage appearing at the input of the apparatus.

The recommendation of BS6651 is that the earthing should be routed as shown in Fig. 3.6b.

While this scheme will protect the line interface circuitry of the apparatus from excessive voltages, it does not protect the power supply of the apparatus, and it is quite possible that insulation breakdown may occur, especially in the mains transformer. The best overall solution is to adopt the scheme of Fig. 3.6b, but to ensure that the earth wire c – e is kept as short as possible.

Protective earth (PE) and functional earth (FE)

A protective earth (PE) connection is only required between those items of mains-powered equipment that require such a connection and the consumer's main earthing connection of the installation. This connection will normally be made via the mains plug and socket. As such, it is a function of the mains electrical wiring and is outside the scope of this book. Note, however, that a test of the integrity of this connection was an important part of the pre-connection inspection (PCI) that, until recently, was required

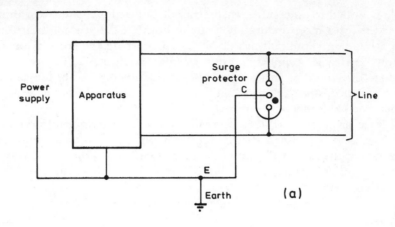

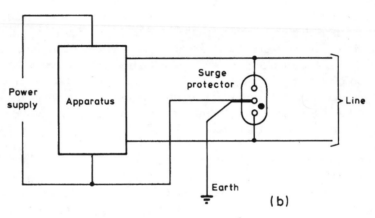

Figure 3.6   *Surge protector earthing.*

for most business systems, and unsafe electrical wiring led to a PCI failure. (The PCI was mandatory until September 1996 when the requirement was discontinued.)

A functional earth (FE) connection only needs to be installed if protective devices, such as surge protectors, have been installed, or where it is an essential part of the installation of a business system where it may also be used for signalling purposes. The wire used for providing and distributing the FE is CW1044, with a minimum cross-sectional area of 1.5 mm². It has a cream-coloured sheath, and is marked with the words 'TELECOMMS FUNCTIONAL EARTH'. The FE connection is made to the consumer's main earth terminal of the electrical installation, and it must not use the consumer's wiring to provide this connection.

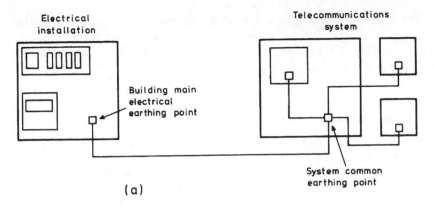

(a)

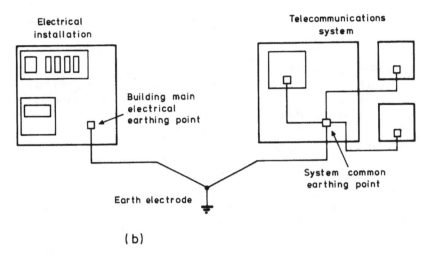

(b)

Figure 3.7 *Provision of functional earth.*

As an alternative, the FE may be connected to an earthed electrode, but this electrode itself should be connected to the main electrical earthing point (see Fig. 3.7).

At whichever of these two points the FE is connected, a label with the words 'TELECOMM EARTH DO NOT REMOVE' should be permanently attached.

Certain multi-pair telecommunications cables contain an FE wire, which is a convenient method for distributing the FE to distribution points away from the main installation. However, if a system extends to more than one building, BS6701 recommends that a separate FE connection be made in each building and that the FE connection should not be carried through inter-building cabling.

It is only necessary to use 1.5 mm^2 wire between the main system installation and any distribution points. The FE wiring to individual stations may use wire of a smaller diameter and will usually be carried within the normal socket wiring.

Cable terminations – sheath removal

The correct method of removing a length of outer sheath of CW1293 and CW1308 cables will ensure that the insulation of the individual wires is not damaged. Firstly, using a sharp blade, 70–100 mm of the sheath are removed. A thin 'rip-cord' will be found laid into the cable, and this is separated out and then pulled back to cut through the sheath as far as required. The sheath can then be pulled away from the wires and cut off and the wires shortened so that any damage caused in the first operation is removed. Figure 3.8 shows the steps involved.

The handling of CW1378 dropwire is not quite so straightforward. Firstly, because of the integral steel support strands, it should be cut only with wire cutters that are specified for use with steel wires. Secondly, the outer sheath is considerably tougher than that of CW1308 cable, and although the same method of stripping can be adopted, it will be found to be rather more difficult. There is a wire-stripping tool that has been designed especially for this type of cable and makes the job much easier (see Chapter 9).

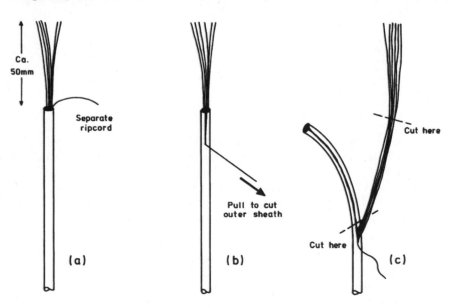

Figure 3.8 *Removing outer sheath of CW1293, CW1308.*

Cable attachment

Virtually all sockets and other interconnection devices have provisions for clamping incoming cables at the point of entry. Either screw-down clamps or nylon cable ties may be used for this purpose. In all cases, the clamping device should be placed over the outer sheath, and it is usually found to be more convenient to attach the cable in this way before the electrical terminations are made.

Wiring of LJUs

When terminating cables, it is important to consider the possible need for maintenance in the future. For this reason, wires should never be reduced to the minimum length; rather, sufficient spare should be left to permit three new terminations to be made. In addition, unused wires in multi-pair cables should not be cut back, but left at the maximum length and folded or coiled neatly. Care must be taken, however, to ensure that none of the wires can become trapped by screws or covers. A suggested method of meeting all requirements when wiring LJUs is shown in Fig. 3.9. Here, the loops of spare wire are formed around the line socket so that on completion, all loops are clear of the mounting screw holes.

Wiring of block terminal

Terminating within the smaller space of a block terminal 77A and similar units is a little more difficult, particularly as the central mounting pillar and screw hole must be kept clear. A stripped cable length of 80–90 mm will be found to be the most appropriate.

Wiring of distribution frames

The term 'distribution frame', strictly speaking, applies to large metal mounting frames, perhaps 2 m in height, which support connection strips and other devices, and cater for several thousand pairs of wires. Distribution frames like this will be found only in the largest of customer installations and in the telephone exchanges of the PTOs. However, the term has passed down to include much smaller structures housed in connection boxes.

Figure 3.10 shows the internal arrangement of a 251 connection box. Boxes of this type are of modular construction – the required number of connection strips type 237 are obtained separately, together with jumper rings, cable fixings, and other accessories. In this example, five 237 strips are fitted to give a total capacity of 50 wire pairs. The 237 strips are snap-fitted to a metal back mount frame, which also provides earthing

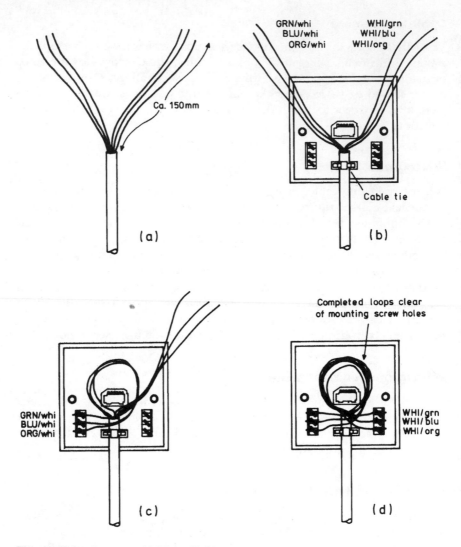

Figure 3.9   *Suggested wiring of LJUs.*

connections to the strips where required. It is therefore important that the frame itself is connected to the functional earth via the supplied earth post and lead, type 5A. This post may be fitted in one of four locations within the box, and is shown at bottom centre in Fig. 3.10.

The jumper rings type 37 are shown in their usual locations towards the corners of the box. They are of a plastic construction with splits at the top to allow the wires to be routed through the rings more easily.

Cable fixings are available in different sizes to suit different cable types and

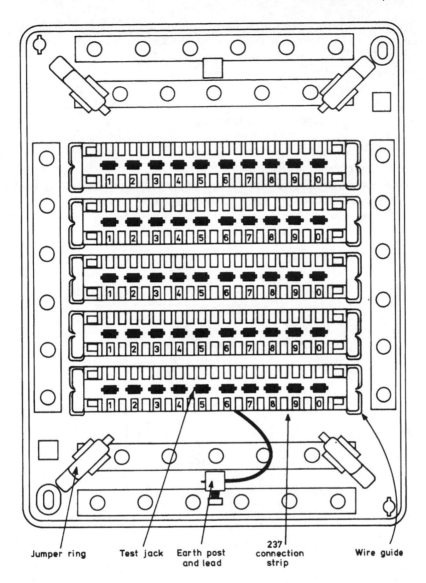

| Jumper ring | Test jack | Earth post and lead | 237 connection strip | Wire guide |

Figure 3.10 *251 box – internal arrangement.*

are fitted in one or more of the holes shown around the outer edges of the box. Cables may be secured to the fixings either by compression-type saddle clamps or by cable ties.

If a larger distribution frame is required, the larger 301 box (100 pair capacity) may be used, and even larger assemblies constructed by bolting together two or more 301 boxes.

Extension wiring from the exchange will normally be routed beneath the connection strips, through the ring on the underside of the appropriate strip, and then fanned out to the terminals along the upper side of the strip, using the wire retainers provided.

Wiring from distribution points and sockets is routed via the jumper rings, and wire guides at the ends of the connection strips, before termination on the lower row of the terminals.

Jumper wiring is then installed to interconnect the exchange wiring with the distribution cables, again routed via jumper rings and wire guides – from the lower row of terminals of the exchange side to the upper row of terminals of the distribution side. Small amounts of excess wire may be conveniently tucked between the terminal strips, and unused wires are usually bundled together and laid neatly up one side of the box.

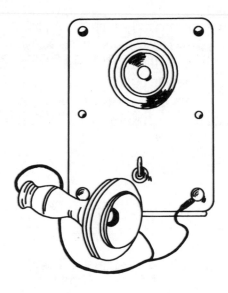

'Butterstamp' U.S.A. 1878

4

Domestic systems

In general, the terms 'domestic' and 'business', as applied to telephone systems, are used to describe the size rather than the application. In practice, a large household could well benefit from a 'business' system, and a very small business might function perfectly satisfactorily using a 'domestic' arrangement.

Domestic systems fall into one of two categories:

- Simple extensions – one or more extensions connected directly to an exchange line.
- Single-line exchanges – a simple private automatic branch exchange (PABX) through which the various extensions may make and receive outside calls, and also make and receive internal ('intercom') calls.

Simple extensions

Basic connections

Once a master socket has been installed by the PTO, connection of any number of extension sockets is very simple, particularly if an NTE 5 type of master socket has been provided.

Each secondary socket has six terminals, numbered 1–6, and the master socket has a similar number. Connection merely entails connecting all of the sockets in parallel – 1 to 1; 2 to 2; 3 to 3 etc., as shown in Fig. 4.1.

If the IDC type of termination is used, note that a maximum of two wires may be connected to each terminal, so that this arrangement is limited to a 'daisy chain'.

It may be that a T-junction is required, such as when taking the lines to the upstairs rooms of a house; the line may emerge centrally, there being one room to the left and another to the right. At this point, a block terminal

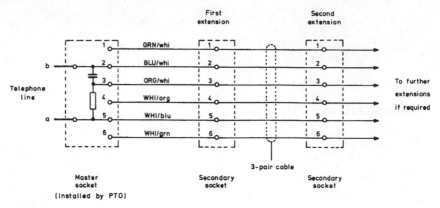

Figure 4.1 *Simple multiple extension wiring.*

77A should be fitted, as shown in Fig. 4.2, thus avoiding the need to branch at a socket, which would require three wires per terminal.

The type of wire used should be either CW1308 for normal work, or CW1316 for under-carpet use.

The colour coding has already been explained in Table 2.4, and assuming that the recommended three-pair cable is used, only the first three elements of Table 2.4 are utilized. The conventional connections are shown in Table 4.1. If, for some reason, only two-pair cable is available, the connections to pins 1 and 6 should be omitted.

Loft wiring

The loft area of premises often provides a useful distribution method. Since the loft floor area is usually used for electrical wiring, it is better to carry telephone wiring as high as possible. This will also help to avoid damage to cables, especially where the loft is used for storage purposes.

The cable is run up a rafter from the point of entry, along the ridge, and down the nearest rafter to the point of exit. It is a good idea to form a loop of spare wire, say about 250 mm diameter, where the cable is closest to the centre of the loft. If further extensions are required in the future, the loop can be unfastened and cut, and the severed cable reconnected using a box connection 77A, which will then provide a teeing point for up to two more extensions.

If the incoming telephone line already passes through the loft on its way to the master socket, one may be tempted to break into the line to provide an extension. This approach must be avoided because (a) it is illegal and (b) if a fault develops, there is no convenient method of disconnecting the extension wiring so that the PTO can test the line up to and including the master socket. The correct method is to provide a second

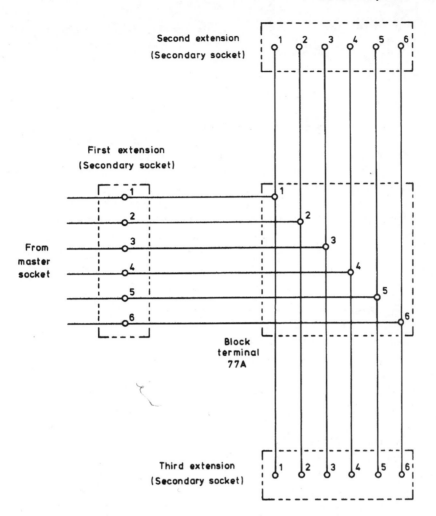

Figure 4.2 *Use of 77A.*

Table 4.1 *Connections to LJU*

Terminal No.	Wire colour	Function
1	GRN/whi	Not used
2	BLU/whi	Line, b-wire
3	ORG/whi	Bell wire
4	WHI/org	Earth
5	WHI/blu	Line, a-wire
6	WHI/grn	Not used

cable from the master socket to the loft, possibly by following the same wiring route.

Limitations

Line length

Generally, extension sockets may be linked to the master socket by not more than 50 m of wiring, and where the wiring has more than one branch, not more than 100 m of wiring shall be used overall. Up to these lengths, multi-pair cables may be used to carry extension wiring for more than one exchange line.

However, if necessary, this maximum length limitation may be increased to 250 m between the master socket and any extension, and 250 m overall where the wiring has more than one branch, in the following circumstances only:

- The wire must be of 0.5 mm diameter or greater (CW1308 is 0.5 mm in diameter).
- There must be no 'series-connected' apparatus, such as call-barring equipment or payphones.
- Extension wiring from one exchange line must not share a multi-pair cable with that of other exchange lines.

Ringer equivalence number (REN)

The number of sockets is not limited, but if the amount of apparatus connected to the exchange line is excessive, there may not be sufficient power for all of the telephone bells to ring.

Before the 'New Plan' system of wiring was devised, the method of connecting more than one telephone to a direct exchange line (DEL) was to use four wires for internal domestic wiring that put the speech circuits in parallel and the bell circuits in series. The original bell-type ringer had a resistance of about 1000 Ω, and it was possible to connect up to four ringers in series and still guarantee that all bells would sound.

Unfortunately, this method is rather too complex to implement in the new system, since it needs a switch contact within the sockets to complete the bell circuit through the remaining telephones when one telephone is unplugged.

The new method adopted is to use ringers of approximately 4000 Ω, and to connect them in parallel instead of in series as before. It is now possible to connect up to four standard bells (or equivalent) in parallel.

A telephone equipped with a 4000 Ω ringer is said to have a ringer equivalence number (REN) of one. Telephones, or other apparatus, of newer design with electronic ringers or ring-tone detectors may exhibit a REN value of more or less than one. When connecting several pieces of apparatus to a DEL, the total of the marked values of REN for each item should not exceed four. If it does, one or more of the ringers may not operate.

Unmarked apparatus of BT origin may be assumed to have a REN of one.

Extension bells

Generally speaking, extension bells or other sounders intended for internal use are provided with a lead and plug so that they may be connected to a conventionally wired socket in the same way as any other piece of apparatus.

However, bells for external use, such as the BT type 80D, are supplied without a connecting cable and must be hard-wired using cable suitable for external use. The two-wire connection is typically by means of a 'chocolate block' screw-terminal connecting strip. In this case, the bell must be connected to terminals 3 and 5 of the socket used for the feed. If it is wrongly connected to terminals 2 and 5, the bell coil will complete a DC path, and the line will then appear to be 'engaged'.

Single-line exchanges

In slightly larger-than-average premises, or in the case of a small business with just two or three rooms, a small exchange may be found to be most convenient. The main advantages of using an exchange are that 'intercom' calls may be made internally (and at no cost since the 'outside' line is not in use), and that calls in progress on the outside line may be transferred between extensions.

Single-line exchanges are surprisingly small. As a result of using microprocessor-based technology, they generally occupy about the same amount of space as a medium-sized book. The problem of finding a suitable location is therefore quite minimal. As far as installation is concerned, the job is hardly more difficult than adding extension wiring.

Exchange designations

The fundamental parameters of an exchange are the number of exchange lines that are catered for, and the number of different extensions that may be connected. These two numbers are then usually combined – '104' ($= 1$ exchange line, 4 extensions); '1232' ($= 12$ exchange lines, 32 extensions), etc. – to describe the size of the system.

System design

Electrical design

Figure 4.3 shows the general arrangement of a basic system. Note that the sockets for the extensions will probably need to be of the 'PABX master'

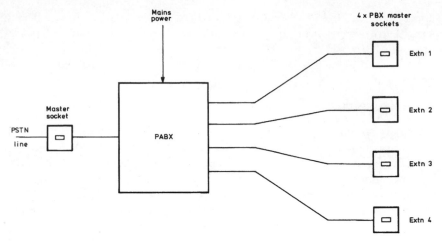

Figure 4.3 *Single-line exchange - basic arrangement.*

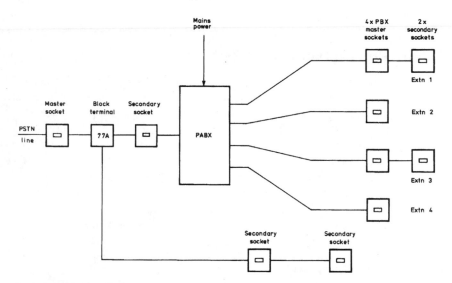

Figure 4.4 *Single-line exchange – additional features.*

type – the variety that contains a bell capacitor. Some exchanges are designed for a three-wire connection – in other words, the bell capacitor is
provided as part of the exchange. The installer then has the choice of
whether to make use of the internal capacitors, making three-wire connections to secondary-type sockets, or to save on cabling costs, making only
a two-wire connection to terminals 2 and 5 of a PABX master socket.

Figure 4.4 shows two methods of expanding the basic facilities. Firstly,
additional apparatus may be connected to the direct exchange line (DEL).

Although such apparatus may be used both to answer incoming calls and to initiate outgoing calls in the usual manner, it is not of course possible to communicate with any of the extensions from these points.

Secondly, each separate extension line may itself be extended to two or more sockets. Note, however, that single-line exchanges are usually severely limited in terms of the number of pieces of apparatus that may be connected to each extension port, and also in terms of the maximum line length that may be connected. This is because these exchanges operate at lower voltage levels than the public switched telephone network (PSTN).

Physical design

When designing the physical arrangement of a system, a number of points should be borne in mind.

Firstly, the PABX will need to be sited conveniently close to both an exchange line socket and a power outlet. It may be necessary to provide one or other of these specially for the purpose, and appropriate steps should be taken to ensure that accidental disconnection of the power plug is unlikely. The site chosen should be both dry and adequately ventilated to prevent overheating.

Secondly, one particular extension number, usually the lowest, will be the designated 'power-fail' telephone. In the event of a power failure, the PABX will cease to function. In order to provide access to the PSTN, in particular for emergency use, one extension will automatically be connected directly to the PTO exchange line, while the other extensions of the PABX will be inoperative. The most appropriate site for this telephone needs to be decided.

Thirdly, system programming will usually be carried out from one particular extension number. This extension may need to be accessible only to authorized persons.

Lastly, convenience of access to the exchange and to any distribution points, both at the time of installation and later, should be considered. This aspect may be made more difficult by the need to make the installation as unobtrusive as possible.

Wiring

Wire type

In most cases, each extension will be wired to the exchange via a separate two- or three-pair cable. Nevertheless, subject to the limitations already mentioned, multipair cables may be used to carry both extension wiring and wiring to extend the exchange line from the master socket. Figure 4.5 shows an example of such an arrangement.

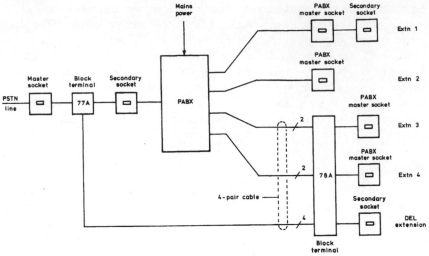

Figure 4.5   *Use of multipair cable to simplify installation.*

Cable identification

There is normally no need for an additional distribution frame at the exchange because of the simplicity of the arrangement, and because suitable terminals are usually provided as part of the exchange. However, to avoid confusion in the future during maintenance or modification, it is recommended that each cable be labelled clearly and indelibly with the extension number and physical destination (e.g. Kitchen, Ext. 2).

If more than one extension is carried within a single cable, then the cable should be identified with the numbers of all the extensions carried.

In order to simplify cable identification, there are various labelling products available. More details are given in Chapter 9.

Wiring practice

Installation of the chosen system will generally follow the pattern described in Chapter 3, with due regard to the manufacturer's specific installation instructions. Note that even though an approved exchange is acting as a barrier between the extension wiring and the exchange line, the same installation practice must be followed as for the simple extension system – otherwise, the approval will no longer apply.

Commissioning

Initial tests

After connecting the PABX to the incoming line, but before applying power to the system, check that the designated power-fail extension telephone gives immediate access to the outside line.

Table 4.2 *Telephone switch settings*

Switch setting	Function
Dialling type:	
T or MF	Tone (DTMF)
P or LD	Pulse (Loop Disconnect)
Recall type:	
T or TB or TBR	Time Break Recall
E or EL or ELR	Earth Loop Recall
Combination switch:	
MF-T or T-TB	Tone dialling + Time break Recall
MF-E or T-E	Tone dialling + Earth Loop Recall
LD-E or P-E	Pulse dialling + Earth Loop Recall
Ringer:	
LO or 1	Ringer volume low
HI or 2	Ringer volume high
OFF or 0	Ringer off

Switch on the power to the system, and ensure that the power indicator light (if provided) is on. Next, establish that an internal dialling tone is available at each extension socket. Then dial into, and out of, each extension in turn. It is easier with two people, using one extension as a 'master'. Make sure that the planned numbering has been achieved, and that all telephones ring when called. As each telephone is connected, check that the dialling mode, ringer, and recall switches (if provided) are correctly set, as shown in Table 4.2.

Telephone settings

Modern telephones are provided with a number of options, usually selected by means of slide switches at the rear, sides, or underneath the apparatus. In some cases, notably some of the newest cordless telephones, options are set by programming. It may be necessary to refer to the manufacturer's instructions. Actual abbreviations tend to differ slightly between manufacturers, but the most common are given here.

It is obviously important that the telephone settings are correctly adjusted according to the requirements of the system to which they are connected.

The type of dialling should be set to TONE/MF unless the PABX in use is an older type and is unsuitable for such operation.

The 'recall' button on the telephone (sometimes labelled simply 'R') is used to recall the PABX dialling tone while placing the current caller on hold. In this way, another extension number may be dialled to enable the call to be transferred to that extension, or for other reasons. Modern PABXs are normally designed for time-break recall operation, and thus the recall selector switch should be set accordingly. If it is wrongly set, the user will be unable to transfer calls.

Time-break recall signals to the exchange by breaking the loop current momentarily. The same effect can be achieved by dialling '1' – as long as the pulse dialling mode is selected. As a further alternative, simply pressing the hook switch (in US parlance, 'flash hook') momentarily will also achieve the desired result. These alternatives mean that telephones without a recall button can still be used.

Earth loop recall signals to the exchange by momentarily connecting the a-wire to earth. In this case, it is necessary to ensure that terminal 4 of the telephone socket is connected to earth. This would normally be carried through the socket wiring, but it does mean that allowance needs to be made in the planning stages for the additional wire.

System programming

Depending upon the type of system in use, a certain amount of programming may be required. This enables various options to be set, and is commonly carried out by making use of a standard telephone connected to the lowest extension number. Some types of exchange have small slide switches ('DIP switches') for setting some options. By whatever means they are set, the options may include some or all of the following:

- Which telephone extensions are to ring on an incoming exchange line call.
- Which telephone extensions are to be allowed to make outside calls – 'call barring'. Note that some systems may arrange call-barring hierarchically; i.e. barring may be introduced between any of the levels Internal/Local/ National/Premium/International However, call-barring at any level should still allow emergency (999/112 service) calls to be made.
- Least-cost routing (LCR) access codes (e.g. Mercury) and associated local code information.

It is essential to make a note of what options have been set, even to the extent of individual keystrokes, so that reprogramming later, if necessary, may be carried out rapidly. This is particularly important if the programmed information will be lost in the event of a power failure, when the user may need to restore normal operation as quickly as possible.

Programming of apparatus

Payphones, equipment with speed-dialling memories, and computers used for internet access that are to be connected to the system, will need to have the exchange line access code (usually 9) programmed, probably including a 'pause' between the access code and the number to be dialled. Refer to the manufacturers' instructions for each piece of equipment involved.

Some specialized items such as alarm systems and electronic banking equipment (credit-card machines) may need to be reprogrammed by their suppliers.

Final testing

Once the system has been programmed, you will need to arrange for an incoming call. (A mobile telephone may help, or there may be another line, perhaps a fax line, that could be used to initiate this call.)

Make sure that all required telephones ring, according to the programming. Answer the call from one extension, and test the ability to transfer the call from each extension to the next until all have been tested. More than one incoming call may be needed to complete this test!

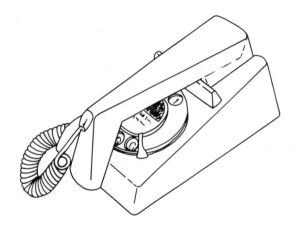

'Trimphone' U.K. 1965

5

Business systems

Types of system

For our purposes, a business system is one that is connected to two or more direct exchange lines ('DELs').

There are several ways in which this may be achieved. The most common approach is to use a PABX, when the incoming lines are connected to a central control unit (CCU) and then routed between a number of extensions. The extensions themselves are able to call other extensions ('internal' or 'intercom' calls), and other features such as call transfer and call barring are normally available. The operation of the CCU may be either analogue or digital, and the CCU may be of fixed capacity (e.g. 308 – three exchange lines, eight extensions maximum) or expandable (e.g. 208 basic, expandable to 416 with additional plug-in modules).

The system may be designed only for use with special proprietary telephones, in which case, it is known as a key telephone system (KTS). Although the special telephones, 'keystations', are well featured, they are more expensive than simple telephones, and since the extension ports are designed for use by keystations only, it is not possible to connect other apparatus such as answering machines, fax machines, or cordless phones directly. For this reason, CCUs are available to which standard telephones and/or compatible apparatus may be connected. Hybrid systems allow for both types of connection. In this case, some extension ports will be designated for keystation operation, and others for standard use.

Since keystation wiring commonly uses all four of the connection terminals in a secondary socket (one pair for voice, one pair for power supply plus data), it is usually called a four-wire connection. A 'standard' connection, using only a pair of wires terminated in a PABX master socket, is called a two-wire connection.

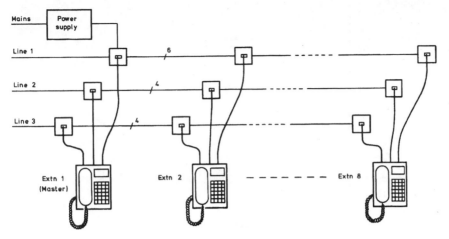

Figure 5.1   *Key telephone system (KTS) typical wiring.*

Alternatively, each port itself might be usable in either mode, determined by the way in which the extension is wired or by some other means such as system programming.

There are some keysystems available that have no CCU at all. The 'intelligence' of these 'multi-line keysystems' is contained within each keystation, and, typically, up to three DELs and eight keystations may be accommodated. Assuming a typical system of this size, each keystation has three line cords, which are plugged into three separate sockets, one for each DEL. This means that all DELs need to be distributed to every keystation position. The connection for Line 1 is slightly different, in that it is a six-wire connection, the additional pair (terminals 1 and 6) being used to carry a low-voltage power supply to each keystation, as well as the intercom. and control signals. The wiring scheme is shown in Fig. 5.1. Extension numbers are allocated by means of programming, and quite sophisticated facilities including hierarchical call barring and ringing assignments may be available.

It is interesting to compare the costs associated with such an arrangement with those of a CCU-based system. From Fig. 5.2, it will be seen that if every station is to be a fully featured keystation, the multi-line keysystem is the more cost-effective. However, if only two or three keystations are required, the other extensions being simple POTs ('plain ordinary telephones'), the CCU-based system is the more economical.

Other possibilities for business systems include cordless systems and Centrex.

Cordless systems offer the ability to communicate with both voice and data, and have the advantage of saving expenditure in rewiring during moves and changes. In a large office environment, this advantage might

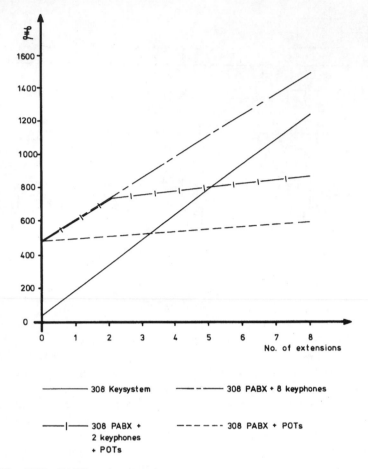

Figure 5.2 *KTS – PABX cost comparison.*

well outweigh the additional cost incurred in providing such a system. Since considerable knowledge of radio-frequency technology is required to plan and implement such a specialized system, it is outside the scope of this book.

Centrex is the generic name for PABX facilities provided by the local exchange. BT offers such a service for large business clients under the name of 'Managed Network Services', which includes data and ISDN access. The advantages of Centrex are that no capital outlay is required to purchase a PABX CCU, and if a customer has two or more geographically different sites, they may be connected to the same system. The disadvantages include the need to purchase telephone sets and wiring, which may approach half the cost of a CCU-based installation, and the payment of an on-going service charge to which the user may be committed for, say, five years.

Hotel systems

The term 'hotel systems' describes a special category of PABX where a number of specialized services are incorporated into the telephone system. A 'hotel package' may be offered as a hardware and software addition to a PABX. In a typical case, the reception desk is equipped with a console, including a QWERTY keyboard and monitor. The receptionist will enter guest registration details, and allocate a room. From this point, 'check-in', the room telephone will be enabled for outside calls, and the room status indicator on the console will show 'occupied'. The guest may request an alarm call, which will be entered into the system for automatic execution at the required time. Any telephone calls made from the room will be automatically billed to the guest.

At the time of check-out, the system will prepare the bill, the room status will change to 'service required', and the phone will be barred from originating outside calls. Once the room has been cleaned, the maid will use the telephone to signal to the system that the room is available again, or that maintenance is required. This information will be displayed on the console, and the receptionist will either re-allocate the room to a new guest or arrange for a maintenance visit. Other facilities, such as baby-listening, may also be arranged through the telephone system. A feature common to hotel packages is that the extension number for each room may be programmed to be the same as the room number. On larger PABX systems, extensions may be designated by 'port numbers' (fixed), rather than by extension numbers (programmable).

Multiple PABX working

It is possible to use more than one PABX, smaller working into larger. This may be a cost-effective alternative to providing a new PABX where the existing system is working at the limit of its capacity, and more extensions are required. It might also be useful where cable capacity is limited, or to provide a small keysystem for manager/secretary use. A general scheme is shown in Fig. 5.3. In this example, a call (internal or external) routed through the main PABX to extension 23 will appear at the secondary PABX as an outside call on line 1. Such a call is programmed to ring extensions 1 and 2. Note, however, that a user in the remote building wishing to make an outside call via the PSTN will need to dial two separate access codes, one to access the main PABX and one to access the DEL.

Other points to bear in mind when considering this arrangement are that with more extensions than the system was originally designed to handle, heavier usage of the DELs may lead to difficulty in accessing an outside line when required; and that compatibility between the existing system and the additional PABX needs to be established. In particular, the REN value of the slave PABX may exceed that of the host.

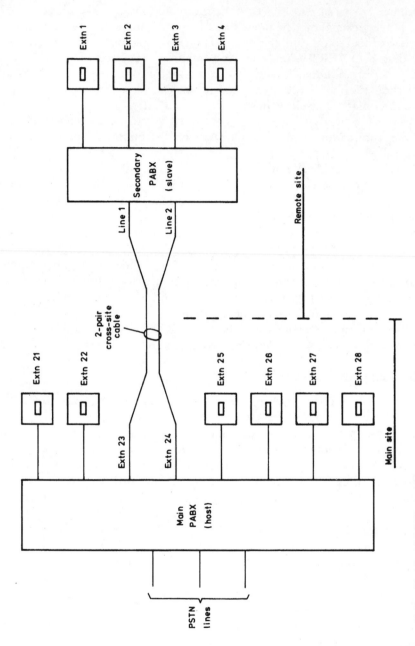

Figure 5.3 *'Behind PABX' working.*

Choice of system

Size of system

Not only should the immediate needs of the customer be considered, but the possible future needs should be taken into account. For example, a small business with two incoming lines and asking for five or six extensions might benefit in the long term by purchasing a 308 system. The incorporation of the fax line into the telephone system is commonly overlooked, but often, a fax line (which of course is no different from any other telephone line) is under-utilized, and if it is brought into use as the primary outgoing line, the voice line is made available for incoming calls. In addition, since the vast majority of outgoing calls, both voice and fax, may be made on the 'fax' line, it may benefit from a high-user tariff, offered in various flavours by BT.

Further savings may be made if the user is a Mercury customer, and a 'Smart Socket' may be installed on the line. (Note, however, that most PABXs now come equipped with automatic least-cost routing (LCR), which obviates the need for additional equipment.)

Features

Many customers are unaware of the facilities available on modern exchanges and may not realize that quite modestly priced systems offer advanced facilities. These include speed dialling (memory dialling with perhaps 100 stored numbers), day/night service switching, automatic call routing dependent upon which incoming line is carrying a call, delayed ringing (if the call is not answered within, say, three rings by the main office, the sales office phone will ring), plus the ability to connect other equipment for paging, music on hold, or building access control. In order to specify the best system for the customer, it is beneficial to gain an insight into the way the customer conducts his business, as well as the geographical layout of the premises.

Hidden costs

When selecting the most suitable system for a particular application, the question of cost will inevitably play a large part in the final decision. One major factor that will affect the cost is whether the PABX will require the provision of an MDF. Some PABXs are supplied pre-wired to a distribution frame, with terminals ready for the connection of DEL and extension wiring, while others are not. Apart from the capital outlay in providing the necessary connection box, the additional labour cost will not be negligible.

If the proposed system is a keysystem, the cost of the special telephone sets (typically between £100 and £150) may easily tip the balance in favour of a PABX that uses standard telephones, and there may also be higher cabling costs to provide four, or even six wires to each extension rather than the two wires required by simple telephones.

Many PABXs include least-cost routing. If the proposed system does not, and the feature is required, some other apparatus will need to be installed – at extra cost.

Optional accessories

Music-on-hold

Many PABXs now offer the facility to provide 'music on hold'. The primary purpose of the feature is to provide the caller with the reassurance that the call has not been cut off, and the source of the music may be either internal or external to the PABX. Internally generated 'music' is usually little more than a simple electronic jingle simulating the chimes of Big Ben, and unless the caller is swiftly rescued, can lead to intense irritation – particularly as the caller is usually paying for the privilege of listening to it. A more pleasing solution is to connect an external music source, such as a cassette tape or CD player, if the system supports the feature. Some systems may need to have an additional module fitted. The programme of music can then be selected to be the most appropriate for the type of caller expected. Thus, a firm of company accountants might opt for a selection of operatic arias, while a surfboard manufacturer would choose to convey a much more up-to-date image. However, the application of music on hold is considered in law to be a public performance, and if the music is copyright, one must first seek the permission of the copyright holder.

Clearly, individual approaches to hundreds of copyright holders are virtually impossible, so the matter is resolved by applying for a licence issued by the Performing Right Society (PRS). A tariff of annual royalty charges applies, which, in 1996 amounted to £66.05 + VAT for a system with up to five DELs, and increasing in stages as the number of DELs increases. Unfortunately, the requirements do not end with the acquisition of a PRS licence. Since the music in this application will inevitably be recorded, a further licence issued by Phonographic Performance Ltd (PPL) will also be required. The 1996 cost for a PPL licence (1–5 lines) was £79.03 + VAT, bringing the minimum annual fee to £145.08 + VAT. (A different tariff structure applies to music-on-hold on freephone/freecall numbers, premium rate services and to local rate numbers. Contact PRS/PPL for further details.) The payment of PRS licence fees is unnecessary if the composer of the music has been dead for 75 years or more, and PPL licensing is required only if the recording itself is copyright. With these in mind, it is theoretically possible to provide music-on-hold quite legally without obtaining either licence, but not without considerable effort and inconvenience.

In order to address this problem, some compact discs have been made available specifically for music-on-hold applications, with the licence fee pre-paid. These are available from Nimans – see Appendix 3.

Paging

In the context of PABXs, there are two types of paging. In the first case, it is sometimes possible to page all extensions, or a selected group of extensions, so that all telephones ring, or a voice announcement made over the internal loudspeakers of keystations. If available, it is an inbuilt function, and requires no extra installation except possibly system programming.

In the second case, the paging, or public address (PA), equipment ('Tannoy') is external to the PABX, and the PABX provides a means of accessing the equipment. This is accomplished by dialling a certain number, as though making an internal call, waiting for a confirmation tone, and then making the announcement. Certain PABXs will require an additional printed circuit board (PCB) or module for this function. Alternatively, adaptors are available to allow a spare extension number to be used for paging.

Building access control

Door phones and door openers are features often available as optional accessories for PABXs. A door phone unit is typically a compact weatherproof unit containing a loudspeaker, microphone and push-button. When a visitor presses the push-button, a certain extension, or extensions, ring with a distinctive cadence, normally three short rings followed by a pause. Which extensions ring is determined by PABX programming. When the call is answered, the visitor is able to announce their identity and nature of business before admission is allowed. The door may be opened manually, or, if a door opener is fitted, it may be released electronically via the telephone system. Systems like this are popular, especially where a receptionist is located some distance from the main entrance, perhaps a flight of stairs away. Provision of these functions may require additional PCBs or modules to be fitted to the PABX.

Printers

Many PABXs support the addition of a computer-type printer for call logging and system management purposes. A low-cost dot-matrix printer, capable of printing on continuous stationery, is usually chosen. The output from the PABX may be either serial (RS232C) or parallel

(Centronics) and the kind of interface needs to be determined so that the appropriate type of printer is chosen.

Headsets

In particularly busy environments, a telephone receptionist might well benefit from the use of a headset – an earpiece/microphone combination. The question of compatibility arises, not only between the headset and the PABX, but also between the headset and the telephone set to which it is connected. Headsets may simply be plugged in, in place of the handset cord of a standard telephone, but there remains the problem of hookswitch control. Others are designed to plug into the headset sockets of compatible telephones. Most major headset manufacturers also market telephones for use with their headsets, but these can only be used on direct lines, and with analogue two-wire PABX extensions. The manufacturers are keen to address the market possibilities of keysystems, and are therefore offering various products to enable their headsets to be used with such systems. Because of the large variety of headsets, telephones, and systems on the market, it is recommended that the advice of either the headset manufacturer or the supplier be sought for each individual case.

Physical installation

Chapter 3 offers guidance in selecting the most suitable sites for the installation of the main equipment (PABX, MDF, etc.) and advice on the location of distribution points (DPs). The MDF, in particular, should be readily accessible and located at a convenient working height. Wiring errors can be made quite easily if it is not possible to view the connection strips 'square on'. For the smaller exchanges, all of the component parts may be conveniently mounted on a wooden panel, which is then screwed into position. This has a number of advantages. Firstly, it means that a large part of the assembly and wiring can be accomplished 'off-site' and in advance – thus reducing the installation time. Secondly, the number of fixing holes to be drilled is reduced to four, and thirdly, it becomes possible to check the basic operation of the system, perform at least some of the system programming, and allow a few hours of soak testing before the actual installation. If the equipment is of an unfamiliar sort, one also has the opportunity to gain some knowledge of the capabilities of the system. A suggested mounting board layout of PABX and associated equipment, based on the Panasonic KX-T 30810/61610 exchanges is shown in Fig. 5.4. Note that all of the wiring between the exchange and NTTA, MDF and programming socket is concealed within the trunking, leading to a very neat appearance.

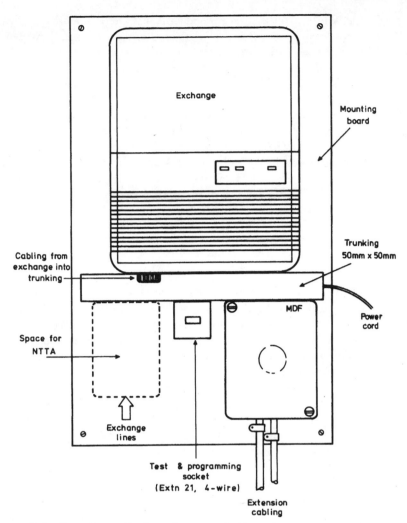

Figure 5.4 *Use of mounting board – suggested layout.*

Wiring

Direct exchange line (DEL) connections

It is most important, at the outset, to establish the boundary between the public network and the user's system. For each exchange line, there will only ever be one demarcation point, and this is called the network termination point, or NTP. Note that the NTP is a concept, not a piece of physical apparatus. The piece of apparatus that forms the location of the NTP is called network termination and test apparatus – NTTA. Individual lines may be terminated at a master socket, as previously described, which then becomes the NTTA. But the master socket is not the only form of termination, and in fact where multiple exchange lines enter premises, they are normally terminated at a distribution frame such as a connection box type 222. This then becomes the NTTA. All the wiring on the network side of this point, including any junction boxes, and the NTTA itself, is the responsibility of the PTO.

In many cases, the NTP will be clearly defined, if master sockets or a connection box already exist. In other cases, it may be necessary to liaise with the network operator to determine the location of the NTP. Where a number of lines are to be brought to a PABX, for convenience and neatness, it is desirable that the NTPs for all of the lines be together and physically close to the PABX. This may mean that the network operator will need to re-route or extend wiring to reach the PABX, for which a charge will be made.

Extension wiring

There are two main approaches to the implementation of extension wiring, and which approach is adopted will depend very much upon the local situation. For the smaller installations, a combination of distribution points with star wiring from those points to the individual sockets will be found to be the most economical and quite adequate. The scheme is shown in Fig. 5.5. The main distribution frame (MDF) is normally located close to the PABX. A number of DPs, perhaps one for each floor of a building, or for other well-defined geographical areas, are connected to the MDF by means of multi-pair cable. In this illustration, part of the MDF has been designated as a DP for the ground floor wiring. An MDF layout suitable for this example is shown in Fig. 5.6. Note that all incoming cables are terminated in a logical sequence on separate connection strips, and the desired connection plan is implemented by using twisted-pair jumper wires. In this way, any extension number can be routed to any location as desired. This can be particularly helpful in the future if there is an office reorganization. Note, also, that the capacity of the cables between the MDF and the DPs exceeds the initial demand. One suggested redundancy figure is to install about three times the number of pairs than the initial requirement, although individual cases may call for a greater or lesser factor than this.

For large systems, where perhaps hundreds of extensions are involved, the 'structured cabling' approach becomes attractive. With this scheme, the DPs are replaced by cabinets containing patch panels – arrays of sockets – which are interconnected according to the current requirements by using patch cords – short lengths of cable terminated in plugs at each end. This enables communication managers to carry out day-to-day changes quickly and simply. Typically, the cabinets are not exclusively for telephone cabling; data network cabling will also be incorporated. The

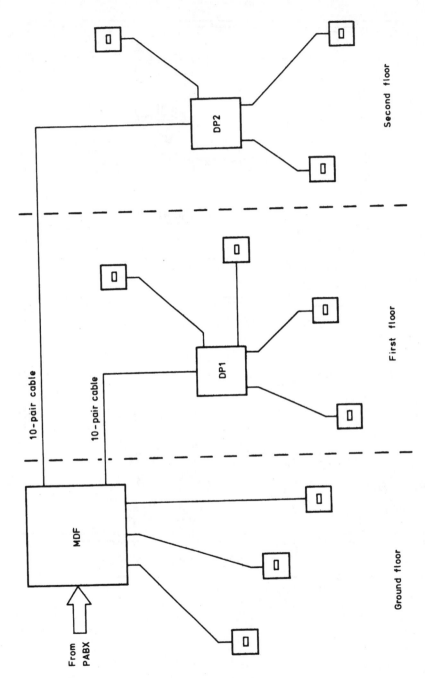

Figure 5.5 *Use of distribution points and star wiring.*

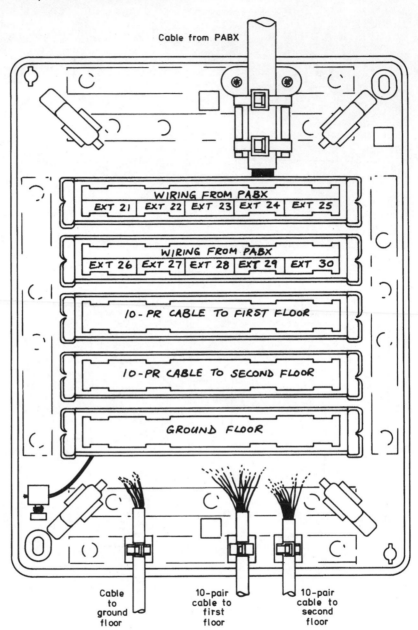

Cable from PABX

WIRING FROM PABX
EXT 21 | EXT 22 | EXT 23 | EXT 24 | EXT 25

WIRING FROM PABX
EXT 26 | EXT 27 | EXT 28 | EXT 29 | EXT 30

10-PR CABLE TO FIRST FLOOR

10-PR CABLE TO SECOND FLOOR

GROUND FLOOR

Cable
to
ground
floor

10-pair
cable to
first
floor

10-pair
cable to
second
floor

Figure 5.6   *MDF layout for example in Fig. 5.5.*

considerable initial expense of installing a structured cabling system will be outweighed by the lower cost of future changes. Chapter 6 explores this topic in greater depth.

Wiring of accessories

Music-on-hold

The connection of a music source to the exchange should be relatively simple, assuming that the exchange manufacturer's instructions are followed. It may be necessary to adjust the sound level after installation.

Paging

The provision of a paging facility may pose a few more problems than music-on-hold. The input to the amplifier from the PABX needs to be kept reasonably short, otherwise background noise, which will be present all the time, not just when the paging system is in use, will become objectionable. However, since the speaker output impedance of most amplifiers will be in the order of 4–$8\,\Omega$, very long runs of speaker lead cannot be tolerated because the losses will be too great. If the length of speaker lead is more than a few metres, an amplifier with a '100 V line' output should be used, together with matching speaker units with integral step-down transformers. By increasing the output voltage, the line current (and therefore losses due to cable resistance) can be reduced, and the speaker impedance is matched at the distant end by means of the transformer. Most speaker units have transformers with adjustable tappings so that the sound level can be adjusted to suit the surroundings of each individual speaker. Care should be taken when siting the speakers. If a speaker is too close to a telephone used to access the paging system, unpleasant 'feedback' or 'howlround' will occur.

In some cases, it may be necessary to install the amplifier some distance from the PABX, in which case, the simplest way of connecting the audio output of the PABX to the input of the amplifier would be by using a spare pair of a multi-pair cable – a balanced line. However, the PABX output and amplifier input will be designed for screened coaxial cable which is an unbalanced line. It will be necessary to use two isolation transformers, one at each end of the balanced line, to make the conversion between balanced and unbalanced. Two identical $600\,\Omega$ audio isolation transformers, available from various component distributors, and suitably enclosed in an insulated protective housing, should be connected as shown in Fig. 5.7.

Access control equipment

Door phone

The installation and connection of a door phone unit should present few problems as long as the unit is compatible with the PABX. The connection

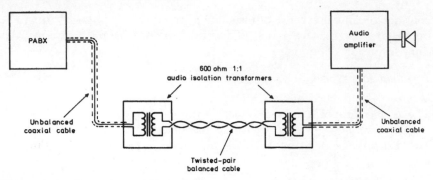

Figure 5.7   *Connection of PABX paging output to remote audio amplifier.*

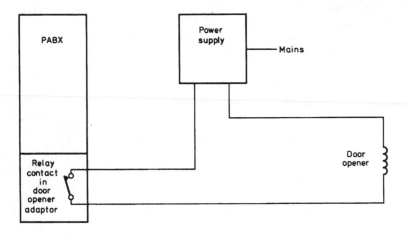

Figure 5.8   *Wiring of door opener.*

is normally two-wire, the unit being powered from the exchange, and the wiring may be included in the same cable as the internal telephone wiring.

Door opener

Door-release mechanisms, of which there are several varieties, are operated from a low-voltage AC or DC supply. Both the mechanism and a suitable power supply will have to be obtained and connected to the PABX, which will provide a timed relay contact closure (see Fig. 5.8). Long cable runs must be avoided, since cable losses may prevent proper operation.

The safety aspect of electronically operated door locks needs to be considered in the context of a power failure. If the safe exit from an area in an emergency is possibly compromised, there are several options. A mechanism that releases the strike plate of a standard 'Yale'-type lock will allow the door to be opened from the inside in the normal way. As alternatives,

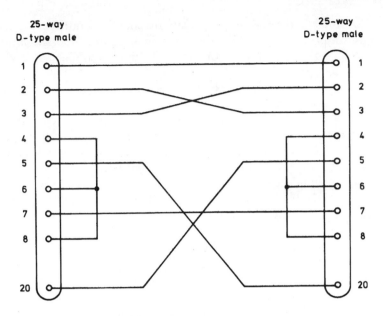

Figure 5.9   *Null-modem cable.*

a power supply incorporating a battery back-up, or a mechanism designed to 'fail open' will allow virtually unhindered passage.

Printers

The printer output from a PABX for call logging purposes may be either parallel ('Centronics') or serial ('RS232C'). In the former case, suitable cables are readily available from numerous sources, such as computer-hardware outlets, and may even be supplied with the printer. In the case of the serial printer, some research may be needed to determine the appropriate connections. In the absence of any information to the contrary, a 'null modem' cable is a good starting point. Details of this cable are shown in Fig. 5.9.

Programming

The first stage of programming will be to reset the system. This may be as simple as a temporary power-down, or it may require a particular programming code to be set. (In the case of Panasonic KX-T exchanges, the code is '99'.) This action should ensure a 'clean slate' before commencing.

Before commencing programming, it is essential to determine with the user exactly what the requirements are. This is especially important if the user will be unable to carry out programming changes himself in the future. The most important areas for discussion will include the ringing, and delayed ringing, assignments, outward dialling assignments, and call barring.

Ringing assignments (i.e. which extensions will ring in response to a call on which incoming line) may differ for day and night service. (Remember that 'night service' may be manually switched in when the telephone receptionist is temporarily unavailable.) Delayed ringing assignments will determine a second group of extensions that will ring in response to an incoming call if the call is not answered within a set time.

Outward dialling assignments for both day and night service will determine which of the outside lines (if any) will be accessible to each of the extensions, and call-barring programming will prevent unauthorized calls being made from certain extensions.

As well as these user-chosen options, a number of other selections and settings will need to be made. These will probably include setting the date and time, outward dialling mode selection for each of the DELs (i.e. tone or pulse – the default setting may be pulse), and PIN-code programming for LCR access, if applicable. Depending upon the particular PABX, the user's requirements and the local conditions, other details may need to be programmed in accordance with the manufacturer's instructions.

In addition to programming the PABX, it may also be necessary to programme any keystation handsets. This should be fairly minimal, normally being restricted to defining the function of a few programmable 'soft' keys, and associated indicators. These are commonly used for 'Direct Station Select' (DSS) purposes. Each 'soft' key is programmed to dial one of the extension numbers, and the indicators show at a glance which extensions are in use at any time.

All programming details should be carefully recorded – manufacturer's manuals commonly include tables for completion (in pencil) for this purpose – and the information kept together with the wiring details shortly to be described.

Documentation

For all but the very simplest of wiring schemes, the need for full and accurate record keeping is readily established. Effective documentation – properly presented and easily understood – can save a great deal of time when the need for repairs or modifications arises. In addition to the programming details already mentioned, there are three areas that should be covered by the documentation – the overall plan, labelling, and wire lists. Probably the most useful, and most often overlooked, is the overall plan. This is a single-page, pictorial representation of the installation. We have already seen a typical example in Fig. 5.5.

Each termination point should be clearly labelled with the same designation used in the overall plan and wire lists. Self-adhesive write-on labels, perhaps also printed with the installer's name, might be used. Alternatively, one of the many labelling products on the market will provide a neat solution.

The MDF should be identified as such, and each distribution point labelled on the cover as 'DP1', 'DP2', etc. Care may need to be taken, if there are two DPs in close proximity, to ensure that the removable covers are not confused in the future. One way of doing this is to position cable entry cut-outs differently so that the wrong covers cannot be inadvertently put in place. Each socket outlet should also be labelled with the extension number and any other important information, such as whether the outlet is intended for a two-wire phone or proprietary keyphone, or if the outlet is meant for a fax machine.

The terminal blocks in the MDF should be labelled – again, in pencil – using the labelling strips available for the purpose. Typical labelling is shown in Fig. 5.6.

While the overall plan gives a general view of the whole installation, in much the same way that a motoring atlas will start with a map of Great Britain showing motorways and trunk routes, it does not carry sufficient detail for individual circuit identifications and test. For this purpose, a wire list should be prepared for the MDF and each DP. Blank forms may be easily prepared and photocopied in advance, and the details entered in pencil so that future changes can be made neatly. The wire list is a record of the source, destination, insulation colour, and purpose of each wire, listed terminal by terminal. For our example of Figs 5.5 and 5.6, there will be one page of the wire list for each of the five blocks of the MDF, plus one page for each DP. Fig. 5.10 shows the wire list for MDF block 3 (MDF/3), and Fig. 5.11 is the wire list for DP1.

The full documentation package, including programming details, should be kept 'on-site' at all times. One way of ensuring the availability of the documentation in the future is to store it in an otherwise empty large connection box, or inside the PABX equipment housing – taking care not to obstruct any ventilation holes.

| Customer / site : | Jones & Co - Green St. Depot | | | | | |
</br>

Customer / site : Jones & Co - Green St. Depot

MDF / Distribution Point

Designation: MDF   **Block:** 3   **Location:** Main office

No.	INCOMING			OUTGOING		
	Wire colour	Cable size	Extn no.	Wire colour	Cable size	Destn
1	Jumpers			whi / blu		
2		From	22	Blu / whi		
3		MDF/1	(4-wire)	whi / ory		
4				Ory / whi		
5			23	whi / grn		
6				Grn / whi		To
7			24	whi / brn		DP1
8				Brn / whi	10 pair	
9			25	whi / gry		First
10				Gry / whi		floor
11			Spare	Red / blu		
12			"	Blu / Red		
13			"	Red / ory		
14			"	Ory / Red		
15			"	Red / grn		
16			"	Grn / Red		
17			"	Red / brn		
18			"	Brn / Red		
19			"	Red / gry		
20			"	Gry / Red		

Figure 5.10   *Example wire list for MDF/3 (Fig. 5.5).*

Customer/site: Jones & Co - Green St. Depot							
MDF/Distribution Point							
Designation: DP1		Block: 1		Location: 1st Floor Landing			
	INCOMING			OUTGOING			
No.	Wire colour	Cable size	Extn no.	Wire colour	Cable size	Destn.	
1	Whi/blu	↑	}	Whi/blu	}		
2	Blu/whi		} 22	Blu/whi	} 3 pr	M.D.	
3	Whi/org		(4-wire)	Whi/org			
4	Org/whi		)	Org/whi	}		
5	Whi/grn	10-pair	} 23	Whi/blu	} 3 pr	} Mb's	
6	Grn/whi	from	}	Blu/whi	}	} sec.	
7	Whi/brn	MDF/3	} 24	Whi/blu	} 3 pr	} Sales	
8	Brn/whi		}	Blu/whi	}	} office	
9	Whi/gry		} 25	Whi/blu	} 3 pr	(Sales	
10	Gry/whi		}	Blu/whi	}	(office	
11	Red/blu		Spare				
12	Blu/red		"				
13	Red/org		"				
14	Org/red		"				
15	Red/grn		"				
16	Grn/red		"				
17	Red/brn		"				
18	Brn/red		"				
19	Red/gry		"				
20	Gry/red	↓	"				

Figure 5.11 *Example wire list for DP1 (Fig. 5.5).*

6

Telephones and structured cabling systems

What is structured cabling?

Background

The concept of structured cabling has emerged over the last few years to address the difficulties that were being encountered in industry. Many different methods of interconnecting computers and peripherals had come into existence that required the use of particular 'topologies' (connection schemes – see Fig. 6.1), and specific cable and connector types, which probably required loyalty to one supplier since different systems were incompatible. In addition, some systems, particularly those using coaxial cable, were intolerant of a problem at one workstation, which could cause a system-wide failure. The expansion of a system often led to different network types existing side by side, which greatly added to the problems of a network manager, who probably had to look after the telephone system as well.

And so the concept was born – one universal cabling system that could be used for practically all existing and foreseen networks, and for any other wired services except for power distribution.

Standards known as EIA/TIA 568A in the United States, and ISO/IEC IS11801 in Europe, have defined customer premises cabling systems for use in medium and large offices. In these standards, several categories of balanced twisted-pair cable have been specified, only two of which need concern us here.

Category 3 (Cat 3) cable, of which CW1308 is an example, is regarded as only suitable for voice, and low-speed data networks up to 10 Megabits (Mbit)/s (1 bit = 1 binary digit).

Category 5 (Cat 5) cable, or Enhanced Category 5 (Cat 5E), is preferred for new networks as it is capable of operating at least to 155 Mbit/s.

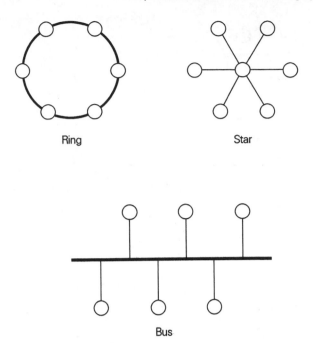

Ring   Star

Bus

Figure 6.1 *Network topologies.*

Much of this improvement in performance has been made possible by improved manufacturing techniques, and by increasing the number of twists per unit length, which reduces very significantly the amount of external noise that is coupled into the circuit, which includes crosstalk. (Crosstalk is the unwanted introduction of a signal from one circuit into another.)

Overall structure

The EIA/TIA 568A Standard has identified six subsystems within a structured cabling system, shown in Fig. 6.2. The entrance facility, (1), is the point where external services such as the PSTN, and private voice and data circuits, enter the building and are terminated. The equipment room, (2), houses the primary communications equipment, such as a PABX, file servers and routing equipment, and maybe data transmission equipment to interface the building system with the external services. Fig. 6.2 shows a number of facilities in the equipment room by way of example. Flexible interconnection is made possible by the use of a patch panel.

Backbone cabling, (3), provides a data 'highway' between the equipment room and localized telecommunications closets, (4). The backbone may be an electrical – 'copper – connection, or by fibre-optic transmission, or by a wireless system. The essential point is that the backbone will be able to

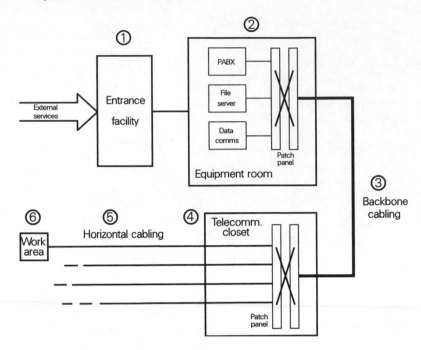

Figure 6.2   *Composition of structured cabling system.*

carry traffic between a number of users and the various services available simultaneously, and will thus be a high-capacity (= high data rates, high bandwidth) link. Backbone cabling is also used to interconnect telecomms closets.

At the telecomms closet, flexible connections are made to the end users, or 'work areas', (6), by means of the patch panel and the horizontal cabling, (5). It should be recognized that the term 'horizontal' refers to the concept that while telecomms closets are likely (but not necessarily) to be located on different floors of a building, requiring cabling between them to be mostly 'vertical', the cabling from a closet to the various work areas will probably be on the same floor, and therefore largely 'horizontal'.

A telecomms closet is also likely to contain items of data comms equipment such as hubs and routers, in order to amalgamate multiple users on to the high-speed backbone cabling.

The work areas (6) consist of nothing more complex than a socket outlet, or, more likely, several socket outlets, allowing several pieces of equipment such as computers and telephones to be connected.

It becomes evident that since a socket outlet in a work area may be connected, by means of patching, to any service, it should not be considered as a data point or as a telephone point, but as a flexible connection point.

While this six-subsystem model is able to help describe the largest system, it is too complex for much smaller systems. For example, a particular system may not warrant more than one patch panel, there may be no identifiable backbone cabling, and work areas may be connected directly to the patch panel in the equipment room. Similarly, the entrance facility is often combined within the equipment room, and the demarcation point will be difficult to identify.

Cable

The most popular form of Cat 5 cable is unshielded twisted-pair (UTP) cable. This four-pair cable consists of the same four coloured pairs as those used in our familiar telephone cable – blue, orange, green, and brown; each paired with either a plain white, or with white/blue, white/orange, white/green and white/brown, respectively. The pairs are enclosed within an outer sheath, usually grey or beige in colour, which is marked sequentially with a length measurement. By reading the measurements at each end of an installed cable, it is possible to determine the length of the run, and it is also possible to determine the amount of cable remaining in the box in which it is supplied.

Another form of Cat 5 cable may be encountered – a shielded twisted pair, or 'STP'. While this shielded cable potentially offers a higher degree of immunity from external sources of interference due to the grounded shielding, it is not often specified for three reasons. Firstly, the cable itself, the connectors and the patch panels are more expensive to purchase. Secondly, it is more difficult to terminate, therefore adding to the cost of the installation. Thirdly, if the ground connection becomes detached at one end, the shield can then act as an aerial, leading to a worse performance than UTP. It is therefore rarely encountered except in military and other secure establishments where the shielding may be required to contain the signal rather than to protect it from outside influences.

Connectors

The designated connector for Cat 5 use is the RJ45, an eight-terminal connection illustrated in Fig. 6.3. Like the familiar 431A plug, the RJ45 plug is an IDC connector that is crimped in a similar way with a dedicated tool. The body of the connector is clear, which enables a check to be made that the wire positions are correct, both before and after crimping.

The socket may be one of a number of types, dependent upon the manufacturer. Faceplates are normally 85 mm × 85 mm, with one or two outlets. A typical example of a single outlet is shown in Fig. 6.4. Some sockets are of modular construction, where the socket modules and faceplates are separate. One socket with blanking plates, or two sockets, may

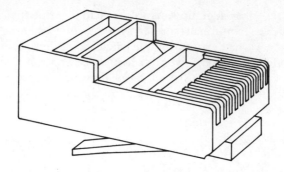

Figure 6.3   *RJ45 plug.*

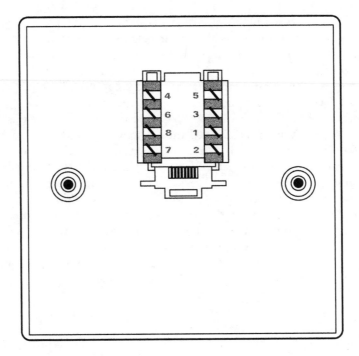

Figure 6.4   *RJ45 socket.*

be snap-fitted into the faceplate. Steel faceplates are also available for fitting the modules into floor box outlet points.

The method of termination is also by IDC, but two differing types are in common use. One type is the same as that used in telephone applications ('Krone'), and the other, slightly smaller, is known as the '110' and requires a different punch-down tool.

Installation

For telephone installation purposes, it is only necessary to consider horizontal cabling, with the terminations at the telecomms closet and at the work area. Backbone cabling and data transmission equipment are outside the scope of this book.

Cabling

The preceding description of the structured cabling concept might lead one to believe that it is something of an overkill when it is used for telephone services distribution. The telephone installer might simply be making use of an existing system or might become involved in a new installation, or an expansion. Because of the flexibility of the system, it is important to realize that today's telephone circuit, with relatively low technical requirements, may become tomorrow's data circuit, and that the installation therefore needs to meet the Cat 5 standard. For this reason, a number of methods and techniques that are quite acceptable in Cat 3 telephone cabling are no longer permitted for Cat 5 installations.

In order to maintain the high-performance capability of Cat 5 cable, the integrity of the twists of the pairs must be guarded. This means:

- There must be no sharp bends.
- No staples must be used.
- The cable must not be crushed.
- There must be no joins in the cables.
- There must be no kinks in the cables.
- The cables must not be pulled too hard (25 lbf maximum is specified).
- Any cable ties used must not deform the cable.
- Direct attachment of cables to surfaces is prohibited. (In many cases, attachment is unnecessary where roof voids and risers can be used, but elsewhere, cables are contained within suitable trunking, ducting, or conduit.).

All of these measures are designed to prevent the pairs from becoming untwisted and/or separated, which would lead to an increase in induced noise, particularly crosstalk.

The Cat 5 specification also includes a constraint on cable length for horizontal cabling. The limit is 90 m, with 10 m allowed in total for patch leads at each end. The imposition of this limit primarily safeguards against excessive attenuation, or signal loss.

For the technically minded, the two parameters of noise (mostly crosstalk) and attenuation both increase with increasing frequency of operation and with cable length. There comes a point where the signal (wanted) and noise (unwanted) become inseparable. The Cat 5 standard requires a minimum margin of 3 decibel (dB), as shown in Fig. 6.5.

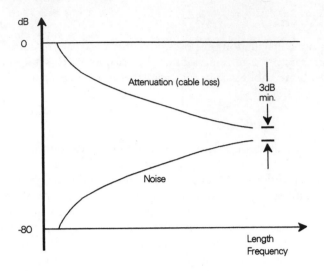

Figure 6.5   *Cable loss.*

Having taken due note of the requirements, the installation of the cable is quite straightforward. A route is determined that is clear of electrical distribution and various mechanical and environmental hazards. Note that a fair proportion of installation time may be expended, together with costs of materials, if trunking has to be installed. It is normally preferable to determine routes and to install all containment (i.e. trunking, ducting, or conduit) before the cables are put into place. If a number of cables are to be pulled at the same time, each cable must be uniquely identified with a number on a label (special labels are available). Care should be taken with numbers that read differently upside down. '68' should be underlined so that it cannot be confused with '89'. At the same time, each box in which the cable is supplied is identified with the same number. It is usually helpful to bind the ends of the cables together with electrical tape. When the cables have been pulled into place, they are numbered again at the box ends (taking the numbers previously written on the boxes) and cut to a length that allows for final routing, termination, and service spare.

Termination

There are several different wiring schemes in use. Details of three are given in Fig. 6.6, but only one is in common use – the '568B' standard. The others are not recommended for new installations but may be occasionally encountered in existing systems.

In all but the smallest installations, termination of each cable will be to a patch panel at one end, and to a socket outlet at the other.

Pin	(Normal) EIA/TIA 568B	EIA/TIA 568A AT&T 258A	USOC RJ45
1	WHI/org	WHI/grn	WHI/brn
2	ORG/whi	GRN/whi	WHI/grn
3	WHI/grn	WHI/org	WHI/org
4	BLU/whi	BLU/whi	BLU/whi
5	WHI/blu	WHI/blu	WHI/blu
6	GRN/whi	ORG/whi	ORG/whi
7	WHI/brn	WHI/brn	GRN/whi
8	BRN/whi	BRN/whi	BRN/whi

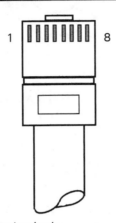

Figure 6.6   *Cable termination standards.*

At the patch panel, the cables may enter from the top or from the bottom of the cabinet. Sufficient cable should be allowed for top-entry cables to reach the floor of the cabinet and then to rise to patch panel height; for floor entry, a coil of spare cable amounting to about the height of the cabinet will remain in the bottom before rising to the patch panel. This amount of excess allows for retermination in the future, the moving of patch panels within the cabinet if required, and, importantly, the temporary removal of a patch panel to a convenient working height and distance for termination and subsequent maintenance.

It is usually easier to terminate the panel outside the cabinet, on a convenient table perhaps, and to dress the cables when the patch panel is put into place. Attention to precise cable lengths at this point will help towards a neat and tidy installation.

Different manufacturers of patch panels have adopted differing styles of cable entry and attachment, upon which will depend the precise method

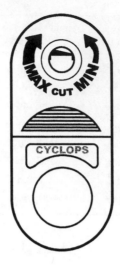

Figure 6.7   *Sheath removal tool.*

of cable routing and dressing. Once this has been established, the outer sheath of the cable is removed to expose the twisted pairs. A simple tool for cutting the sheath is available, shown in Fig. 6.7. The cable end is inserted into the cutter, and the entire tool is rotated around the cable to score the sheath. One direction of rotation produces a deep score that may well damage one or more of the wires inside, while the other direction produces a shallow score that does not penetrate the sheath completely. To avoid problems later, always use the shallow direction, marked on the tool.

The sheath must be maintained as close to the termination point as possible and each pair untwisted back only as far as the termination point. No more than 13 mm may be separated after the surplus has been removed. Using a suitable punch-down tool, the wires are terminated – the colour code is provided by the patch panel manufacturer. Once all the cables have been terminated in their appropriate places, cable ties are used to secure the cables – not too tightly – to the panel rear, preventing any strain being taken by the terminations. The weight of the cable loom on the patch panel is then relieved by attachment down the inside of the cabinet with cable ties.

Outlet sockets

Much the same procedure is followed as with the patch panel, although cable dressing is much simplified. Sufficient cable is left coiled inside the socket housing to allow for up to two new terminations, and sheath stripping and termination follow the same rules as before, although the terminal

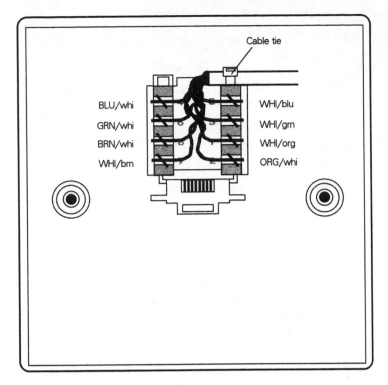

Cable tie

BLU/whi

GRN/whi

BRN/whi

WHI/brn

WHI/blu

WHI/grn

WHI/org

ORG/whi

Figure 6.8 *Terminated RJ45 socket.*

layout will be different. Fig. 6.8 shows the rear of a typical terminated socket.

RJ45 plugs

Very small installations may not warrant the use of a patch panel, which is a fairly expensive item. An alternative approach is to terminate the cables with RJ45 plugs at the 'patch panel' end. These will then be connected directly to the hub.

The method of termination is similar to that described for the 431A/631A telephone connector, although it is a little more tricky as the wires need to be untwisted and arranged in order before insertion into the plug. For the smallest installation, where two computing devices are connected directly together, a 'cross-over' lead may be needed. Details are shown in Fig. 6.9.

Test and measurement

Once the cables have been installed and terminated, they are ready for testing. The simplest form of test is a continuity check, and a number of

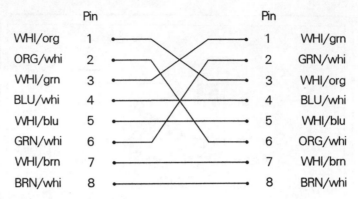

Figure 6.9   *Cross-over lead.*

two-part testers with LED indicators are available, costing about £100. In use, one half of the test set is connected to the workstation outlet socket, and the other half to the appropriate patch panel socket. In order to make these connections, two patch leads will be required that obviously will need to be 'known good'. Once connected, the tester automatically tests each wire pair sequentially for continuity and for crossed connections. Make sure that the LEDs on both halves of the tester give correct indications – sometimes, a fault will be indicated at one end only. This variety of tester is capable of identifying most cabling faults.

More advanced, and more expensive, measurement devices are also available. These will check continuity, crosstalk, attenuation, and length under simulated operating conditions, and will allow results to be stored for subsequent print-out via a compatible computer program. Some measurement sets also incorporate a time-domain reflectometer that enables pinpointing of cable faults.

Labelling

To complete the installation, each outlet needs to be labelled with a unique identifier corresponding with the patch panel. It may be that a simple, sequential numbering scheme will suffice, taking advantage of the manufacturer's numbering of the patch panel outlets. If more than one cabinet is involved, then outlets may be numbered 1/01, 1/02, etc., then 2/01, 2/02, etc., where the first number is that of the cabinet, and the second is the patch panel outlet number.

Use of structured cabling for telephone connections

So far, we have only considered the requirements of Cat 5 cabling without regard to its intended usage. That is as it should be, since Cat 5 cabling is intended as a universal connection scheme. In order to use this distribution structure specifically for voice communication, we need to consider the method of connecting a telephone extension to the horizontal cable at the patch panel end, and how to connect a telephone at the workstation.

For convenience and simplicity, one or more rows of the patch panel will be dedicated to telephone extension outlets. This section of the patch panel will therefore be hard-wired back to the exchange, normally using the standard wire colours. Extensions may be two-wire, utilizing the blue pair, and others may be four-wire, using the blue and orange pairs, or possibly the blue and green pairs. In any case, not all of the eight connections for each patch panel outlet will be used. If there is any doubt as to which connections to use, it would be sensible to connect only one extension and test it before proceeding with the rest of the terminations. Since the cabling between the exchange and the patch panel is hard-wired, voice-grade Cat 3 cable of any convenient size may be used.

A short Cat 5 patch lead connects the desired extension patch panel outlet to the required horizontal cable leading to the workstation.

At the workstation, an adaptor will be necessary to convert from the RJ45 outlet socket to a 431A socket into which a telephone may be plugged. There are three types of adaptor – master, PABX master, and secondary – which perform the same functions as the similarly named sockets. A PABX master adaptor will be required if a standard telephone is to be connected, and a secondary adaptor is used if the telephone is to be a proprietary four-wire instrument. The completed wiring scheme is shown in Fig. 6.10.

There is a fourth adaptor, called an ISDN adaptor, that is used when an ISDN-compatible device is to be connected. Unlike the other adaptors just mentioned, the output socket is also RJ45, and the connections are 'straight through'. However, termination resistors are incorporated that ensure that signals are not reflected from the distant end of the cable, that could cause problems leading to data corruption.

The flexibility of the system can now be fully appreciated. Assume that Joe is assigned to workstation 7, and can be reached at telephone extension 25. Due to his dedication to duty, he is promoted to become head of section, and may thus enjoy the privilege of a private office. By simply moving the patch lead across from position 7 to position 2, workstation outlet 2 is now connected to extension 25, and Joe may keep his telephone number, which is known to everyone. The important point here is that the wiring change is accomplished without having to call in an engineer to move jumpers on a distribution frame – it is simply done, by almost anyone, by plug and socket.

Before leaving the subject of structured cabling, it would be as well to mention another handy little device – the two-way splitter.

It will have been noticed that while a Cat 5 cable consists of four pairs,

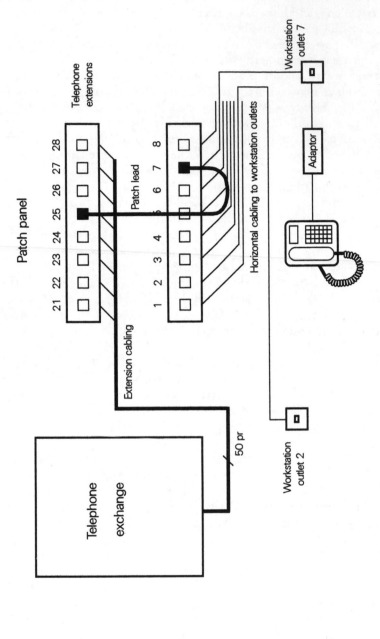

Figure 6.10 *Structured telephone cabling.*

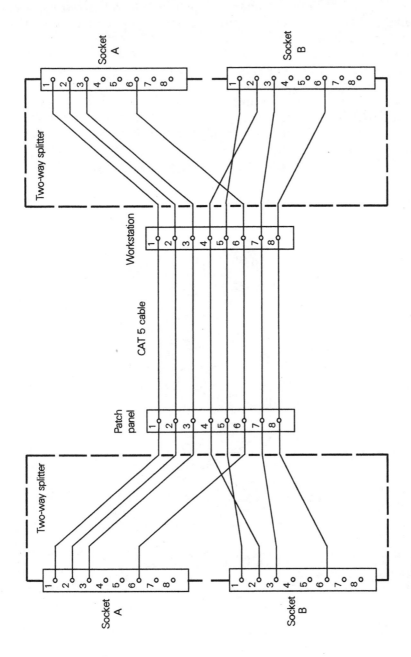

Figure 6.11 *Two-way splitter wiring.*

only one or two of the pairs are used for telephone work. (Indeed, only two pairs are used for the most common computer links – a transmit pair and a receive pair.) This means that the spare pairs could be pressed into service if two circuits are required where only one cable is available. The two-way splitter, one of which is installed at each end of a horizontal cable, enables this to be done simply and quickly, without disturbing the structured cabling. A diagram is shown in Fig. 6.11. While more than adequate for voice communication purposes and low-speed data links (i.e. 10 MHz or less), these two-way splitters are likely to cause degradation to high-speed data links, and their use should therefore be restricted to non-demanding applications.

'Desk Set' U.S.A. 1897

7
Fault-finding and prevention

Fault-finding

The logical approach to fault-finding involves determination of the fault symptoms, the formation of a diagnosis based upon observation of the symptoms and the results of tests and measurements, and the correction of the problem. In most cases, the existence of a fault is confirmed by the appearance of one or more symptoms, the nature of which will determine the steps to be taken. Usually, a particular symptom may have a number of causes. For instance, the absence of dialling tone when one is expected may be due to a broken wire leading to the socket (i.e. an installation fault), or to a faulty earpiece in the telephone (an apparatus fault). For this reason, fault symptoms in the following category of 'Installation faults' may in fact be due to an apparatus fault, and vice versa. However, it was felt to be important to preserve the concept of fault identification by symptom rather than adhering strictly to the nature of the fault.

Tables 7.1–7.5 summarize the installation fault-finding section and are included as a quick reference guide.

Installation faults

Symptom – no dialling tone

This symptom may be due to one of four causes:

- wiring
- termination
- apparatus
- service.

Table 7.1   *Fault-finding chart – 'dead' line*

Symptom	Possible cause	Action
No dialling tone	(1) Break in wiring	Locate and rectify or replace
	(2) Short in wiring	Locate and rectify or replace
	(3) IDC connector faulty	Locate and replace
	(4) Too many wires in IDC terminal	Rearrange – max. two wires per terminal
	(5) Different-sized wires in IDC terminal	Use different type of termination
	(6) Apparatus fault	Locate by substitution and replace
	(7) Fault on incoming line	Check at master socket, then refer to PTO
	(8) Power to PABX turned off (PABX only)	Restore power

Table 7.2   *Fault-finding chart – ringing problems*

Symptom	Possible cause	Action
No ring on incoming call	(1) Break in wiring (especially to socket terminal 3)	Locate and rectify or replace
	(2) No master socket	Install master socket (PABX only) or refer to PTO (direct lines)
	(3) Faulty socket	Locate and replace
	(4) Too many telephones connected in parallel (REN exceeded)	Reduce number of telephones in parallel
	(5) Apparatus fault	(1) Check ringer switch 'on' (2) Check by substitution, then replace
	PABX installations only:	
	(6) Programming error	Check and rectify
	(7) 'Do not disturb' set	Clear DND
	(8) 'Call forwarding' set	Clear
	(9) Telephone incompatible with PABX	Replace with compatible type
Continuous ringing when one handset lifted	(1) Wiring incorrect (a / b wires reversed between sockets)	Check and rectify
	(2) 'a' wire broken	Check and rectify

Table 7.3 *Fault-finding chart – dialling problems*

Symptom	Possible cause	Action
Unable to dial	(1) Telephone set to wrong dialling mode	Check and set to correct mode
	(2) (PABX only) Wrong dialling mode set automatically	Turn off PABX power, restore power, reprogramme as necessary
	(3) Incoming line set to LD (or MF) only	Set telephone accordingly, or refer to PTO
	(4) Incoming line has permanent bar on outgoing calls	Refer to PTO

Table 7.4 *Fault-finding chart – call transfer problems*

Symptom	Possible cause	Action
Unable to transfer calls	(1) Operator error	Check method being used
	(2) Telephone set to wrong recall type	Check and set correctly
	(3) Earth connection missing from socket terminal 4 (Earth Loop Recall only)	Check wiring and rectify

Table 7.5 *Fault-finding chart – degraded service*

Symptom	Possible cause	Action
Audible hum on line	(1) Unbalanced line caused by faulty apparatus	Check by removal or substitution and replace
	(2) Unbalanced line caused by wiring fault	Check and rectify
Crackling on line	(1) Intermittent connection or short circuit	Locate by isolation, then rectify
Interference	(1) Crosstalk	Separate different services into different cables
	(2) Radio interference	Install filter(s) Install screened cables Coordinate with radio operator and/or Radiocommunication Agency

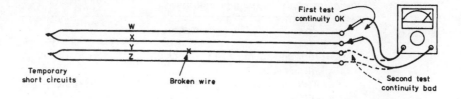

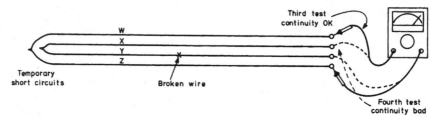

Figure 7.1   *Continuity testing for suspected broken wire.*

In new installations, the most likely cause of a problem is a wiring error, which may well be revealed by a careful check at all junctions and terminations.

Wiring problems may arise either during or after installation. Where a staple gun has been used to secure the cable, a staple may have punctured the cable and either shorted together two or more wires, or severed one or more wires. Continuity testing should reveal if either of these conditions exist. In Fig. 7.1, a broken wire is identified by two simple tests. In the first test, both wires labelled W and X are confirmed as good. In the second test, the W and Z wires are confirmed as good, leaving only the Y wire as faulty. To investigate a suspected short circuit condition, the temporary short circuits of Fig. 7.1 are removed, and continuity tests from each wire to every other wire of the cable are carried out.

Once a cable fault has been identified electrically, the cable should be inspected to try to ascertain the cause of the problem. It may be that it is being subjected to constant abrasion, has become trapped by a door or window or furniture, is being regularly trodden on, or is attacked by rodents. Ideally, the entire length of cable should be replaced, taking due care to re-route or protect the cable against a recurrence of the problem. For particularly long or awkward runs, the damaged section may be cut out and replaced, using approved jointing methods. As a temporary measure, and where there is spare capacity available, the service may be transferred to another wire pair on the same cable, with the faulty wire(s) being identified at each end to prevent future use.

Termination problems may arise with insulation-displacement type connections for a number of reasons. The type of wire may be unsuitable for the terminal, or there may be wires of different diameters sharing the same terminal. In some cases, the terminal itself may be damaged, especially if an attempt has been made to make a connection using an unsuitable tool such as a screwdriver. This is likely to force apart the two sides of the contact so that the insulation is no longer pierced when the wire is inserted. A wire may not have been punched down sufficiently firmly, or there may be too many wires in one terminal – two is the maximum permissible.

Apparatus faults are best identified by substitution. Some common and correctable problems with apparatus are dealt with in the next section.

Service faults may also give rise to a lack of dialling tone. If the installation is a simple extension or extensions to a direct exchange line, there may be a line or other fault that is the responsibility of the PTO. If there is still no dialling tone at the master socket when all the extension wiring has been disconnected, then the fault should be referred to the PTO.

If the faulty extension is connected to a PABX, check first that power is applied to the PABX. Remember that one or more extensions will provide direct access to DELs in the event of power failure, so the presence of dialling tone on one extension is not necessarily proof that power is applied to the PABX.

Symptom: no ring on incoming call/unwanted ringing

Ringing problems may be categorized in one of five ways:

- wiring
- termination
- apparatus
- configuration
- service.

Most of the problems of wiring and termination have already been discussed, the only difference being that the bell-wire connection will be the area requiring attention.

One fault symptom, which may be initially puzzling, is unwanted ringing – when one handset is lifted, other telephones on the same circuit ring continuously. This is due to a combination of factors:

- Ring generator circuits in modern telephones will operate quite satisfactorily with a DC applied voltage as with AC – including the DC exchange line voltage.
- When a telephone handset is lifted, a low-value (about $100\,\Omega$) resistor is connected between terminals 3 and 5 – across the parallel bells. This is intended to damp out bell tinkle due to pulse dialling.

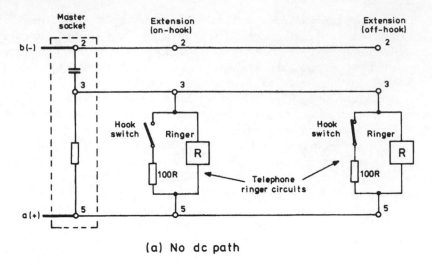

(a) No dc path

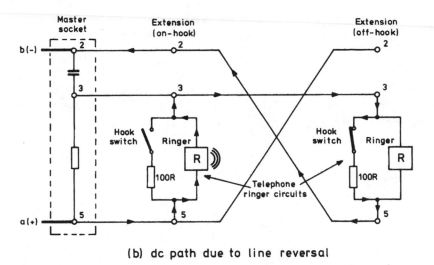

(b) dc path due to line reversal

Figure 7.2   *Unwanted ringing due to line reversal.*

● One of two possible faults will now cause unwanted ringing. Firstly, if the a and b wires are inadvertently reversed in the extension socket wiring, a DC path is established through the ringer circuit of other telephones still 'on hook' (see Fig. 7.2). Secondly, a break in the 'a' wire (to socket terminal 5) will allow a DC path to be established, as shown in Fig. 7.3. The cure, of course, is to locate the line reversal or break and correct it.

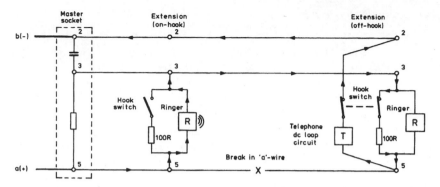

Figure 7.3 *Unwanted ringing due to break in 'a'-wire.*

The lack of a master socket in a circuit will of course prevent any telephones from ringing. This should never arise on a DEL as the master socket will already have been fitted by the PTO, but secondary sockets may have been inadvertently fitted to PABX extensions. Occasionally, faulty sockets may cause the symptom. Chemical deposits, possibly formed by condensation from drying plasterwork, may cause problems by causing corrosion or alternative conductive paths.

Ringing problems will also result from having too many pieces of apparatus connected to the same circuit and the REN value being exceeded. Many small PABXs have very limited REN values, and in some cases, this may be as low as one.

Apparatus faults leading to ringing problems may include faulty ringer circuits, or parallel connected telephones being 'off hook', or appearing to be 'off hook' because of a fault.

Configuration problems may be due to switch settings or programming faults in PABX installations. Most telephones have ringer on/off switches that may have been accidentally set to the wrong position.

In PABX installations, misdialling from an extension may cause 'do not disturb' or 'call forwarding' to be programmed, both of which will cause an apparent ringing fault. Similarly, PABX system programming can prevent certain extensions from ringing on incoming exchange line calls.

There may occasionally arise a problem of incompatibility, particularly if old-style telephones with mechanical bells are connected to small PABXs. In some cases, the ringing voltage in PABXs is derived from the AC mains and is at the mains frequency of 50 Hz. Electromechanical bells are unable to respond to this frequency, being designed to operate at 16–25 Hz. In other cases, the ringing waveform itself may be unsuitable for specific telephones. Early models of the BT Diverse 1000 cordless telephone do not respond to square-wave ringing signals found on several small PABXs, for instance.

Symptom: unable to dial

The most common cause of this symptom is that an incompatibility exists between the dialling mode expected and the dialling mode produced. PSTN lines from modern exchanges should be able to accept both pulse and tone dialling, but individual lines may be set for pulse dialling only. Dual-mode dialling can be requested of the network operator as long as the local exchange is suitable. Certain PSTN lines may have been set, at the request of the subscriber, to bar outgoing calls. This limited form of service is sometimes requested by landlords of rented accommodation. Obviously, for dialling to succeed, the bar would have to be removed, with the permission of the landlord.

Modern PABXs can normally accept both dialling modes, but in the special case of the MDS Opera exchange, each extension port has the ability to 'learn' the type of dialling automatically. On first power-up, each extension is set for pulse or loop disconnect dialling. If a tone-dialling telephone is connected, the first time it is used, the exchange will switch automatically to tone (MF) mode. If, subsequently, a LD telephone is now connected, the exchange will not switch back to MF, and it will be impossible to dial. This can cause problems if telephones are swapped around when pulse-dialling-only telephones apparently cease to function. The cure is to remove the power to the exchange for a few minutes and then reapply it when the default conditions will be restored.

Symptom: unable to transfer calls

In order to transfer an outside call from one extension of a PABX to another, the recall facility must be used, as described in Chapter 4. Many problems arise through 'operator error', and thus, if difficulties are being experienced, the procedure being followed should be checked as the first stage. The next most likely cause is that the type of recall selected by telephone switch setting is incorrect – check that it is set to 'time break' or 'earth loop', as required by the PABX. In the case of 'earth loop' recall, the earth connection to terminal 4 of the socket should be checked.

Symptom: degraded service

There are three main areas where faults may lead to degraded service. Firstly, a high level of hum, caused by an unbalanced line, has already been described in Chapter 1. Although the most likely cause is a wiring fault, the hum may also be introduced by a piece of faulty apparatus. As this possibility may be eliminated easily by disconnection, it is the logical first stage in fault diagnosis.

Secondly, other noise of a more random nature such as crackling, is most likely due to an intermittent wiring fault. Damaged cables and corroded terminals may not be the easiest to locate, especially if they are underground, and it may be felt that replacement of the affected line in its entirety is the more cost-effective approach. There is a chance that a piece of apparatus may cause the difficulties, so this possibility should be eliminated first.

Thirdly, interference of several forms can cause a degraded service. The mechanism by which unwanted signals are picked up is quite complex, but in all cases, the level of interference is altered by the distance between the source and the cabling that is affected. Crosstalk is the term used to describe the unwanted introduction of signals from one circuit to another, and is usually due to the two circuits sharing the same cable. In many instances, simply splitting two circuits into two separate cables can reduce the crosstalk to an acceptable level. In more persistent cases, it may be necessary to find a different route for one of the circuits.

Sometimes, telephone wiring may be susceptible to interference from strong nearby radio signals such as taxi services. Some benefit may be gained from installing a line filter – a modified version of the 80A connection box designated '80A RF2', or it may be necessary to replace internal wiring with screened cables, where the screen is earthed at one end. In extreme cases, a degree of cooperation will be necessary with the operator of the transmitter concerned, and possibly with the involvement of the Radiocommunication Agency. (Contact information will be found in Appendix 3.)

Apparatus faults

In this section, only faults on apparatus that may be rectified without intruding into areas that are not intended to be accessed by the user will be considered. Other faults may well require a high degree of electronic expertise and access to sophisticated test equipment and are outside the scope of this book.

Chapter 8 provides an insight into the principles of operation of the major types of telephone apparatus and may also prove helpful in fault diagnosis.

General

Line cords are probably more susceptible to damage than any other part of a piece of telephone apparatus. They are frequently trodden on, tripped over, trapped in doors, and chewed by dogs. Happily, they can be repaired and/or replaced quite quickly. There are one or two points to remember, though, when replacing the 431A plug on leads of apparatus of foreign origin. Firstly, it may be found that the D-shaped cordage has only two cores rather than the expected four. If this is the case, the two wires should be spread so that they enter the channels of the outermost contacts of the four-way plug. In order to ensure a satisfactory termination, it may

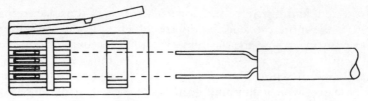

Figure 7.4   *Termination of two-wire line cord.*

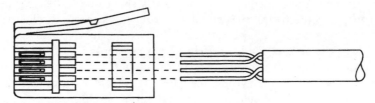

Figure 7.5   *Termination of certain line cords (e.g. Panasonic).*

be necessary to strip back a little more of the outer sheath than the normally recommended 15 mm (see Chapter 2 and Fig. 7.4).

Secondly, at least one manufacturer, Panasonic, has a rather mischievous way of line cord termination. Because the 'a' and 'b' wires are carried along the middle two wires of the cord instead of the outer two, the wires have to be reorganized, as shown in Fig. 7.5, before crimping. This is a rather fiddly operation, and one or two attempts might be needed before success is achieved. As it is impossible to tell visually how a 431A plug has been wired, a continuity test of the original cord and plug may need to be carried out to be certain.

Handset cords, by the nature of their use, are also vulnerable items. If the cord is of the type with connectors at each end, replacements in different lengths are widely available for user replacement.

Payphones

Payphones probably suffer from more problems than any other piece of telephone equipment because they can undergo more than usual physical abuse. Mostly, damage of this type is easily seen, but may or may not be easily remedied.

Accidental or deliberate damage aside, coin jams or other coin-chute problems probably account for the majority of faults. Such faults are usually indicated by the message '999 calls only' and may simply be due to a full cash box. Optical sensors are often used in coin discriminators, and any obstruction of an optical path will be identified as a coin jam. Removal of foreign matter may require a certain amount of dismantling and reassembly.

Other payphone problems are commonly due to flat batteries, which in turn may cause loss of programmed information, including the clock. As the payphone will not be able to determine the appropriate charges to be made, '999 calls only' is, again, a likely response. In some cases, batteries are used to operate coin-release solenoids, which require high currents for short periods, and payphones of this type should only be fitted with alkaline cells or the life expectancy of the batteries will be unacceptably short. When batteries have been replaced, the programming should be checked and restored if necessary.

Some payphones are fitted with microswitches that are actuated by the case-lock mechanism. Opening the case signals a different mode of operation, such as a 'programming' mode, and if the microswitch fails to operate properly, difficulty will be experienced in normal use. In some cases, lock-switch failure will allow only three digits to be dialled, after which the line connection is broken. The cure is to adjust the mechanism until the microswitch actuation is positive.

Answering machines

Answering machines will fail to function correctly if the tones recorded on the tape, by the machine, to mark the start and end of messages are not recognized. There are two main reasons for this failure – a dirty record/playback head or a worn tape. In the first case, the head can be cleaned with a cotton-tipped applicator moistened with a solvent such as alcohol ('meths' is fine), and in the second case, the tape may be turned over so that the other end is put into service (remembering to re-record the outgoing message), or simply replaced.

Malfunctions may also occur if the drive belt of the tape mechanism stretches and then begins to slip. This should be obvious by observing the movement of the tape, and the cure for this problem is the same as for a broken drive belt, which is replacement. Although not really a 'user-accessible' repair, the job requires only a little common sense and dexterity, and a few small tools.

Occasionally, a tape cassette is inserted incorrectly so that the tape forms a small loop around the rear of the capstan spindle instead of passing directly between the spindle and the pinch roller, as shown in Fig. 7.6. The motor and capstan operate, but the tape does not move. This problem should become obvious on close inspection, and the remedy is to remove the cassette, tension the tape in the cassette, and re-insert.

Cordless telephones

A poor operating range with cordless telephones is often attributable to broken or shortened aerials. A broken handset telescopic aerial is easily detected, and almost as easily replaced. An abnormally short wire aerial from

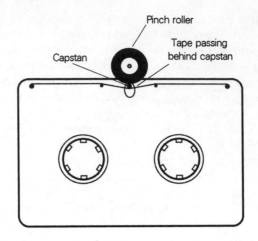

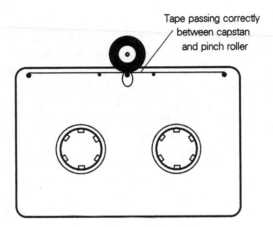

Figure 7.6   *Incorrect cassette insertion causes malfunction.*

the base unit may not be quite so obvious. Unfortunately, there is no 'standard' length for these aerials, which differ between models, and may be anywhere in the range of 1.9–3.0 m. If a 'known good' example is not available for comparison, it is suggested that a damaged aerial lead should be lengthened to about 2.5 m.

Handset battery packs have a life expectancy of between 1 and 2 years, depending on how well they have been treated. Occasional complete discharge and recharge cycles will help to prolong their life. Most modern handsets are equipped with plug-in battery packs in their own compartments, and replacement takes a matter of a few seconds. Older models require the entire case to be opened, and sometimes the battery pack connections may need to be soldered. Prices of replacement battery packs can vary widely – trade suppliers offer packs in various sizes for £3 or so, while some retail outlets may charge ten times this figure, so some shopping around may pay dividends.

Battery-charging problems may be due to dirty or corroded contacts either on the base unit or on the handset. In some cases, this may also lead to a failure to exchange security-code information, which will prevent communication between base and handset being established. More information on this aspect is given in Chapter 8.

Fax machines

Most of the simple problems with fax machines can be accounted for by the accumulation of dirt and debris, or the use of wrong-sized paper rolls.

Dirt, paper fragments and correction fluid deposits in the region of the document reader will typically cause black vertical lines to appear on the copy. Occasionally, a complete document may become wrapped around a roller and cause a paper jam or slippage. Clearing the problem may or may not be simple, depending on the design of the machine.

Other problems arise if the wrong type of paper roll has been installed. Fax-paper rolls come in many different sizes of paper width, core diameter, and roll diameter. If a width of paper is used that is too narrow, it is possible for an edge-mounted paper detector to miss the paper completely – perhaps after a period of trouble-free operation during which the paper drifts to one side – and an 'out of paper' alarm is signalled. Some machines are able to accommodate two different paper widths, but a spacer must be installed for the narrower roll to prevent drift. Other difficulties may arise if the paper roll merely lies in a well, when dirt or sticky deposits may prevent the roll from unwinding smoothly.

Fault prevention

In large part, the requirements and provisions of BS6701 and the Oftel Wiring Code are designed to minimize the likelihood of the occurrence of faults. In particular, BS6701 warns against the installation of wiring where it would be exposed to any of a number of hazards.

Abrasion, and other forms of physical damage, are usually predictable and therefore avoidable by choosing appropriate routing. In particular, external overhead cables that are subjected to excessive movement in high wind conditions need to be positioned to avoid any possibility of contact with roof tiles, guttering, and the like. For internal cables, the use of ducts and conduits will provide a high degree of protection against mechanical hazards.

In extreme cases – for example, heavy engineering workplaces – the use of armoured cable in exposed and vulnerable locations should be considered.

Protection of cables against moisture is well provided for. It is only necessary to choose the most appropriate cable type for the conditions expected. This may mean that for some internal work, external-grade cables would be a sensible choice.

The same argument holds true for connection boxes. Unfortunately, socket outlets are not suitable for external use or for particularly damp conditions. The solution to this difficulty is to install a weatherproof telephone with a hard-wired connection.

Cables, connection boxes, and sockets are suitable for use over a wide temperature range, but clearly, the installation of thermoplastic components either directly, or in close proximity, to heat sources must be avoided. At the low end of the scale, damage to PVC cable insulation is likely to result from attempting to form the cable at temperatures below $0°C$.

Mention has already been made of the potential for damage to sockets from chemical deposits from new plasterwork. As no sockets are available that are adequately protected in this respect, probably the safest course of action is to replace the sockets installed in new plasterwork after, say, six months of use, if no problems have already developed. Perhaps, a manufacturer would care to address this shortcoming, and offer 'tropicalized' sockets for use in new plasterwork and other environments where dampness or fumes prevail.

Rodent damage to wiring may be expected to occur in any location where a food supply exists. The most obvious places are farm outbuildings and areas where waste foodstuffs are stored. The use of armoured cable overcomes the problem.

Items of telephone apparatus can be damaged and abused in a number of ways. For 'rough service' areas, the simple precaution of wall-mounting a telephone will help to prevent damage from dropping and splashing, and wear and tear of the line cord. When choosing a telephone for areas where rough handling and accidental damage are to be expected, it is probably more cost-effective to opt for a low-cost model, and operate a discard-and-replace policy, than to pay a high price for a vandal-resistant type. If the problem really is one of vandalism, then of course the expensive option must be the solution.

Damage caused by lightning strikes continues to plague both users and their insurers. Such damage occurs when the potential difference (voltage) between two points in a system exceeds the value that the insulation between those points is able to withstand, and arcing occurs. Arcing itself is a flow of current, and localized heating may cause the fusing together of the two points that were previously insulated from each other, or elsewhere within the current path that has been established. In some cases, the heating effect will cause an open circuit. Where the lightning strike has been severe, with both high voltages and currents, sufficient power may be dissipated to cause very extensive damage.

Any piece of apparatus that is connected to both the telephone line and to the mains power is vulnerable, because the mains supply is held at or near earth potential, whereas the telephone line is free to adopt any potential.

Any piece of mains-powered telephone apparatus will have a clearly defined boundary between the telephone line and the mains supply. Power is allowed to cross the boundary through a transformer, and signals may cross the boundary through either transformers or optical isolators. Components like these are permitted at the boundary because they do not provide a direct electrical path – there is 'voltaic isolation'. As these components form the 'front line', they naturally tend to suffer first from extreme conditions.

The first line of defence is to introduce an intentional short circuit from each side of the telephone line to earth if the potential difference rises much above the level normally expected. These are the surge protectors that have already been discussed.

Surge protectors act by encouraging arcing in a controlled environment when the potential difference rises above a certain point – around 200 V. This has the effect of clamping the voltage to that level.

More recent thinking indicates that there may be more than one mechanism by which lightning strikes may cause damage, and that the very inclusion of surge protectors may be instrumental in opening a second path. The suggestion is that if lightning strikes the ground at some point, there will be a temporary, and substantial, rise in the potential around that point compared to the general ground potential some distance away. In order to even out this distribution, currents will flow, taking the easiest route, which may be provided by electrical power lines. The entry point to the electrical system could be across a surge protector (in the reverse direction to that expected) and through the telephone apparatus boundary, causing as much damage as necessary in the process. Figure 7.7 illustrates the principle.

If one then also considers the possibility that lightning may strike an overhead mains power line and cause discharge currents to flow across the apparatus boundary to earth via the surge protector, it becomes apparent that there is no sure way of protecting against lightning damage.

Until all our communication lines are optical, the safest course is to unplug all apparatus if there is an electrical storm in the offing.

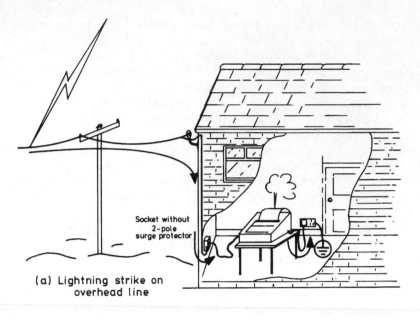

(a) Lightning strike on
overhead line

Socket without
2-pole
surge protector

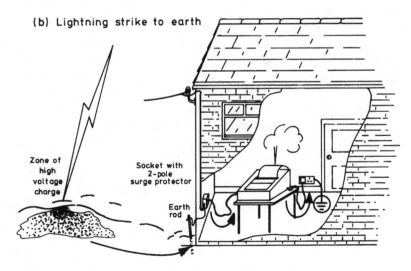

(b) Lightning strike to earth

Zone of
high
voltage
charge

Socket with
2-pole
surge protector

Earth
rod

Figure 7.7   *Two ways to destroy a fax machine.*

8
Apparatus

Wired telephones

The simplest form of telephone, often known as a 'POT' (plain ordinary telephone, or plain old telephone) must incorporate all of the features already described from a technical standpoint in Chapter 1. These features may be summarized as follows:

- a microphone and earpiece;
- a means of completing the DC local loop; this allows signalling to the exchange that either the user wishes to initiate a call, or that an incoming call has been answered;
- a means of dialling;
- a means of indicating the arrival of a call.

To this list of essential requirements may be added a simple memory that enables the last number to be redialled automatically (useful if the number was engaged on the first attempt), and the ability to store a limited selection (usually about ten) of frequently dialled numbers.

These functions, with little or no enhancement, may be built into some decorative housing, such as an ornate reproduction antique with much gilding, or an amusing shape, such as a car or cartoon character. Telephones in this class are known as decorative or novelty telephones.

When these features are added to, the result is the 'feature phone'. Typical features include single-button dialling, where one of a set of buttons may be pressed to dial a frequently called number. Sometimes, one of these buttons will be identified as the 'Mercury' button, and it may be pre-programmed with the user's own LCR access code. Since large numbers of buttons take up a lot of panel space, the effective capacity may be doubled by the addition of a 'shift' key so that as many as 40 different numbers may be stored.

On-hook dialling, or hands-free dialling, allows a user to set up a call without first lifting the handset. The ringing tone and the voice of the called party are relayed over an internal speaker. The handset is lifted when the called party answers to allow two-way conversation.

An enhancement of this feature is called full hands-free operation, where a microphone is incorporated into the body of the telephone. This eases the difficulty of referring to paperwork in an office environment and also allows several people to join in a telephone conversation at once.

The liquid crystal display (LCD) allows the called number to be displayed and, with the addition of an internal clock, may show the time of day and the duration of the current call.

Special-purpose telephones may have particularly large dialling buttons for the partially sighted, or may be 'hearing-aid compatible' for the hard-of-hearing. In this case, the handset contains a coil of wire through which the earpiece current passes, which allows inductive coupling to a hearing-aid pick-up coil.

For those who may spend the greater part of the working day using a telephone, some types have provision for headset use, and for special environments, weatherproof, vandal-resistant and hazardous area models are available.

Wired telephones are sometimes combined with other telephone apparatus such as cordless telephones or answering machines, but these cannot really be considered to be feature phones.

Cordless telephones

Since the mid-1980s, when cordless telephones were made available to the domestic market, their popularity has far exceeded expectations. This has led to the development of standards to permit more radio channels to be used and to improve security.

Before going any further, it is as well to make the distinction between cordless telephones and cellular telephones, especially as both types may be called 'mobile phones'.

The cordless telephone may be regarded as an ordinary telephone that is plugged, as normal, into a standard telephone socket. The handset cord is replaced with a low-power two-way radio link, which allows greater mobility on the part of the user. The range of operation is short – normally up to about 200 m – although some non-approved (and therefore illegal) types exist that have a much greater range than this. Except for these illegal types, they are exempt from licensing by the Radiocommunications Agency, and their use does not incur any higher charge than for the same call made from an ordinary wired telephone.

Cellular telephones, however, are connected by means of a radio link to a network such as Cellnet or Vodafone. Each telephone has its own individual number, and, subject to network radio coverage, may be used anywhere in the UK and, in some cases, in other countries as well. The networks are operated quite separately from the PSTN, but are interfaced to it to enable calls to be made between cellular telephones and the PSTN. Such calls are subject to higher charges than most calls within the PSTN.

As already indicated, the cordless telephone consists of two parts – the base unit and the handset. The base unit is connected to the telephone line, and to the household electrical supply, which provides power for the electronic circuitry. The handset is powered by means of a rechargeable battery pack, which, depending upon the amount of use, will power the handset for some 8–12 h. The battery is recharged by placing the handset on the base unit, which supplies charging current through electrical terminals. Sometimes, these terminals become dirty or corroded, which prevents proper charging. An indicator light is usually provided to show that charging is taking place. Separate charger units are sometimes available, which may even incorporate an alarm clock and broadcast band receiver, suitable for the bedside table.

Each of the handset and base units is a transceiver – that is, each unit transmits a radio signal to the other, and receives a radio signal from the other.

The first standard, called CT1, allows for only eight different radio channels. The base unit transmits at one of eight frequencies around 1.7 MHz, and the handset transmits at one of eight frequencies around 47 MHz. The actual frequencies are listed in Table 8.1.

This arrangement means that the handset may receive the signal from the base unit on a compact internal ferrite rod antenna, similar to the type used in a medium-wave transistor radio. The telescopic aerial, which is used only to transmit to the base unit, may be collapsed when not in use,

Table 8.1 *CT1 Radio frequencies*

Channel number	Transmit frequencies (MHz)	
	Base	Handset
1	1.642	47.45625
2	1.662	47.46875
3	1.682	47.48125
4	1.702	47.49375
5	1.722	47.50625
6	1.742	47.51825
7	1.762	47.53125
8	1.782	47.54375

since it is not necessary to have the aerial extended in order to receive a call from the base station.

With only eight different radio channels available, there is a significant possibility of interference between two different cordless telephones, but the chances of accessing someone else's telephone line either deliberately or accidentally are much reduced by the use of a security code.

Each base/handset pair is preset to a particular code that is transmitted each time the telephone is used. Neither unit will respond to the other unless the correct code is received and recognized. Sometimes, this code may be altered by setting a series of miniature switches, ensuring that the switches are set identically in both handset and base unit. Some later models are arranged to change the code automatically each time the handset is placed on the base unit. Although this scheme effectively prevents others from learning the security code and thus being able to make unauthorized use of the telephone line, it is a mixed blessing. If there is a power interruption to the base unit, the system will cease to work until the handset is replaced on the base unit for the security code to be set up again. In addition, since the code is exchanged between the two units by means of either electrical contacts or an optical path, dirt or foreign matter may prevent successful operation.

A later development of the cordless telephone is the DECT (Digital European cordless telephone) CT2 standard. A much higher radio frequency is used, between 864 and 868 MHz, which makes eavesdropping much less likely. Forty channels are provided for, thus lessening the chance of interference, and a system known as the Common Air Interface, or CAI, has been established. This means that handsets of different manufacture may be used with one base station. Base units are no longer set to fixed channels. Instead channel seeking is used to find a vacant frequency.

It is possible to have more than one handset in use with one base unit, and 'intercom' calls between handsets may be made. In addition, incoming calls may be transferred from one handset to another. A cordless telephone of this type may be considered therefore as an alternative to a small PABX.

The speech is transmitted in digital form between the handset and base unit, which further enhances security and quality. The actual process consists of taking an instantaneous sample of the audio signal, measuring its voltage, and transmitting that value in numerical form. This sample is followed by another, and yet another, at a rate of perhaps 10 000 samples per second, or even greater. The reconstruction at the distant end reverses the process, generating a voltage according to the received value. What the listener hears is now the result of a mathematical process carried out electronically, and not a signal that has suffered the effects of being transmitted by radio. Thus, radio noise is virtually eliminated, and if the radio signal is intercepted, the eavesdropper hears only the buzzing noise of data being transferred.

Features

Some variations between different models of cordless telephones are to be found. In some cases, a means of 'paging' the handset from the base unit is provided. That is, pressing a button on the base unit will cause the handset to emit an audible calling tone that may invite the called party to 'pick up' the current call, or which may be a prearranged signal for some other purpose. Alternatively, this feature may be enhanced by incorporating a full intercom facility, and in the case of multiple handset arrangements, intercom calls may be possible between handsets as already mentioned.

Other facilities may include a memory, a microphone 'MUTE' (or 'SECRECY') button, LCD, and some indication of battery condition for the handset. Some handsets provide an audible warning of a battery going flat, and of the approach of the limit of operational range.

Since the operation of a cordless telephone is dependent upon power being available at the base unit, it is clear that the system will cease to function in the event of a mains power failure. Some models provide for the installation of standby batteries either in the base unit itself, or in the separate mains power supply, to overcome the problem. These will only give a limited period of operation and cannot provide enough power for handset recharging, which is therefore disabled during a power failure.

Some models may be set to operate on one of two different radio channels. If interference is experienced with a cordless telephone belonging to a neighbour, the channel may be altered to overcome the problem by means of a slide switch on the base unit, and/or a channel change button on the handset.

Installation

To achieve the best possible performance from a cordless telephone, the siting of the base unit should be considered carefully. Generally, the chosen location should be as near as possible to the centre of the proposed area of operation. To gain coverage of the widest possible area, it may be beneficial to install the base unit upstairs or even in the loft space. It is worth remembering that each solid wall that the signal travels through will reduce the signal strength by about half. Metal-clad walls and steel structures will cause even more deterioration than this.

The wire antenna should be laid horizontally as far as is possible, and the telescopic antenna should be fully extended in the vertical position. Because of the vagaries of radio propagation, and the possible presence of unseen hindrances to radio communication, such as hidden cables or foil wall insulation, some experimentation in the positioning of the base unit may pay dividends.

Before use, the handset battery will need to be charged fully. Most manufacturers recommend a period of between 10 and 16 h for this. Users will maximize the life of their rechargeable batteries by periodically allowing them to discharge completely, and then recharging them fully again.

Telephone answering machines (TAMs)

Basic functions

Modern technology, in particular the microprocessor, has allowed the complex circuitry of the TAM to be condensed into a remarkably small package.

The most basic TAM must provide all of the following functions:

(a) The owner is able to pre-record an announcement message, usually called the outgoing message, or OGM.
(b) When the machine is activated, it detects an incoming ringing signal, and counts the number of rings.
(c) After a predetermined number of rings (sometimes adjustable), a relay or solid-state switch, is turned ON to complete the local loop, thus simulating the lifting of a handset.
(d) The pre-recorded message is played to the line, usually asking the caller to leave a message after a tone.
(e) The mode of operation is now changed from playback to record, a tone is sounded, and the audio from the telephone line (the incoming message, or ICM) is recorded.
(f) Once the ICM is complete, recording ceases, and the telephone line is released.
(g) This cycle of operations may be repeated many times, with all the ICMs being preserved until the owner is able to listen to the messages.
(h) Once the TAM has answered a call, it is normally possible to intercept the call by picking up a telephone connected to the same line. (The TAM can detect the resulting additional voltage drop on the line.)
(i) When the owner wishes to retrieve the messages, they may be played back, several times if necessary, after which the ICM recording space becomes available for recording future messages.

Microprocessor control

A microcontroller is the most obvious choice in order to control all these operations. A microcontroller is a single-chip microprocessor, where the central processing unit (CPU), read-only memory (ROM), random access memory (RAM), and various input and output (I/O) circuits are combined into one component, or chip. The required operations are pre-programmed into the ROM during manufacture, which means that the device is 'application specific' – it can only be used for the purpose for which the program was written.

The CPU may be thought of as a 'control room'. Messages are received from the outside world – in this case, the ringing signal, button pushes, end-of-tape detectors, etc. – and instructions are issued in accordance with the procedures, or program. Such instructions will be prioritized and will control tape movement, recording or playback, the telecom. line relay and many other features. It may be important to know that a button labelled 'PLAY', for example, does not itself control the 'play' operation. Instead, it signals to the CPU that the operation is required, but the CPU may countermand the instruction if, say, the end of tape has been detected. Fig. 8.1 is a block diagram of a basic microprocessor-based TAM.

Magnetic recording

Where the recording medium is magnetic tape, the answering machine must mark the beginning and ending of messages. This is accomplished by recording short bursts of tone at specific frequencies, which, on playback, the machine can recognize. It is worth remembering that a dirty record/ playback head may cause such a degree of deterioration of the recorded tone that the machine will no longer be able to recognize the signals. By the same token, a worn tape, or stretched and slipping drive belt causing speed, and therefore frequency, variations, may also cause a malfunction.

Digital recording

In order to overcome these problems, TAMs are now appearing on the market with digital recording. This method dispenses with the mechanical tape deck and replaces it with electronic circuitry that converts audio signals into digital form. The conversion process is similar to that described for digital speech transmission in some cordless telephones. In this case, however, the digitized speech is not transmitted over a radio link, but stored in a RAM. On playback, the contents of the RAM are retrieved and converted back to the original audio signals.

Once the messages have been retrieved and are no longer required, the contents of the RAM may be overwritten by further messages.

Added features

Basic TAM specifications may be enhanced by one or more of the following features:

(a) The TAM may be combined with a standard telephone and handset, or, more rarely, with a cordless telephone or fax machine.
(b) Some magnetic tape machines are equipped with two separate tape mechanisms and two separate tapes – an endless loop cassette for the OGM and a standard cassette tape for the ICM. This arrangement has numerous advantages. Firstly, as the OGM tape is an endless loop,

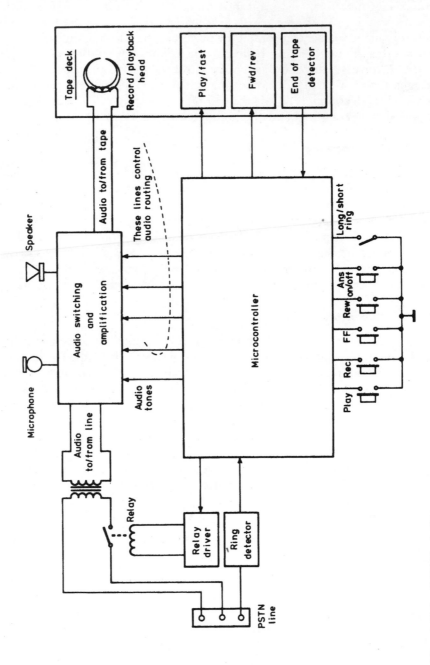

Figure 8.1 *TAM block diagram.*

the deck needs to be able to drive the tape only in one direction and at one speed. The tape is simply advanced until a short metallized portion is detected, when it is stopped, ready for the next time it is needed. Secondly, the ICM tape does not need to be rewound after each message so as to be ready to play the OGM on the next call – the OGM is recorded separately. Instead, the tape is stopped at the end of each message, ready for the next message. This is a useful feature where a large number of incoming calls are expected, since later callers do not have to wait until the tape is repositioned over earlier messages.

(c) A 'MEMO' feature allows a message to be left on the TAM, using the internal microphone rather than the incoming telephone line, in much the same way as the OGM is recorded. In this case, however, the message is retrieved in exactly the same way as a normal ICM. Thus, members of busy households may leave messages for each other with reasonable certainty that they will not be missed.

(d) Where the owner is away from home for some period, some TAMs allow remote operation. The owner calls his home number, and when the TAM responds, he interrupts the sequence by sending a series of tones, usually by dialling a special number code on the keypad of a DTMF telephone. On recognizing the code, the TAM will play back the recorded messages to the telephone line. By using other codes, messages may be saved, or deleted, or the OGM changed if desired. Some TAMs incorporate a 'toll saver' feature. If there is already a message waiting on the machine, it will answer an incoming call within, say two rings. If there are no messages waiting, the caller will hear perhaps six ring tones before the TAM responds. The owner, knowing this, will be able to hang up after three of four rings, without incurring any charges.

(e) Certain TAMs are equipped with an internal time clock which, in conjunction with an electronic voice synthesizer, allows each message to be annotated at the end with a date and time announcement.

(f) When the owner returns, the presence of recorded messages for his attention may be announced by a simple flashing light. More advanced machines will display the number of incoming messages by means of a numerical display.

(g) Sometimes, it may be desirable to choose between two different OGMs, depending upon the circumstances. Some TAMs allow for more than one OGM and even permit the switch to be made remotely.

(h) Occasionally, a TAM may be used only to give information, not to record it, such as an alternative number to ring. Some machines are equipped with a selector switch to choose between 'NORMAL' and 'ANNOUNCE ONLY' operation.

(i) Some people may value their answering machines for their ability to screen calls. It may be preferable to allow every call to be answered automatically, and then to break into the call once the identity of the caller has been established. In this way, unwelcome callers may be ignored, but it does have the disadvantage that other callers may hang up as soon as they realize that their call has been answered by a TAM.

Modems

For many years, the PSTN has been used not only to carry the voice traffic for which it was designed, but also for data transmission. Almost all such transmissions are between computing devices of one sort or another. Practically all computers are digital, where information is converted to binary digital form for processing and transfer. The binary system uses only two digits, '0' and '1'. These two digits are usually represented electrically by 0 V DC , and a low positive voltage, typically 5 V.

There are two methods of transferring data from one point to another – parallel and serial transmission. In the parallel method, many lines all carrying binary data operate in parallel so that, say, 16 bits of data can be transferred simultaneously. This is the fastest means of transfer, but requires a large number of separate lines, which, over long distances, becomes expensive. The second, slower, method is to use one line, and transmit the data in serial form, 1 bit at a time. For a given data rate, this means that the transmission time would be 16 times longer than for a 16-line parallel scheme.

Using a telephone line, only a serial data transmission scheme can be used. Unfortunately, the data stream of alternating DC voltages cannot be transmitted directly by the PSTN, and so a device is used that converts the DC levels into tones. The device, which is also designed to perform the reverse operation, is called a MODEM (= MOdulator–DEModulator). In its simplest form, the modem will generate a tone of one frequency to represent a binary '1', and a tone of different frequency to represent a binary '0'. This is called 'frequency shift keying' or FSK (see Fig. 8.2). More complex schemes than this to increase the data rate are now in common use, and are described in detail in Reference 4 (see Appendix 2).

There are three different modes of transmission:

- Simplex – transmission is only possible in one direction (e.g. radio broadcasts.)
- Half duplex – transmission is possible in both directions, but not at the same time (e.g. walkie-talkie radios.)
- Full duplex – Simultaneous transmission in both directions is possible (e.g. telephone conversations)

A system of operation ('protocol') for modems allows full duplex operation. The originating modem uses one pair of tones, and the answering modem automatically selects a different pair of tones, thus allowing both

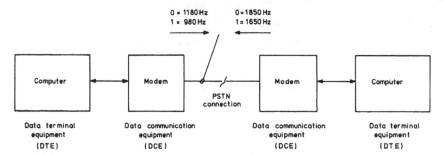

Figure 8.2 *Modem full-duplex operation using FSK at up to 300 bits/s.*

modems to transmit simultaneously without interference. The additional bandwidth required for full duplex operation limits the data rate to 300 bits/s.

Like any other piece of telephone apparatus, modems must carry BABT approval, and the physical connection to the PSTN is exactly the same. Hook switch control, dialling and ring detection may all be carried out by the host computer under program control, so there is no need for a separate telephone to be connected to the line.

Fax machines

Introduction

The facsimile ('fax') machine is playing an increasingly important role in our daily lives. There are two major advantages of a fax message over a telephone call:

- An A4 page of information, which can be hand-written, typed, printed, or drawn, can be transmitted from source to destination in under a minute – much less time than it would take to dictate or describe the same information by voice.
- It provides a hard copy at the distant end, and is acceptable in lieu of a signed letter when placing orders or issuing instructions to banks, etc.

The latter is made possible because of the method of transmission. In the older telex system, now largely superseded, each alphanumeric character was transmitted as a digital code that controlled an automatic typewriter at the receiving end. Any kind of disturbance in the transmission path could cause a character to be changed, thus possibly changing the meaning of a message. Additionally, since only alphanumeric characters could be sent, it was not possible to transmit a signature, which would be needed to make an instruction legally binding.

Operation

The fax machine treats all documents as pictures, regardless of the content. In order to transmit a picture, the page is scanned by a combination of mechanical and electronic means, and the image is broken down into individual picture cells, or 'pixels'. In the simplest case, each pixel is determined to be either black or white, and the appropriate tone signal transmitted to the distant end. Most fax machines are also capable of transmitting different shades between black and white, known as 'grey scale', and thus, it is also possible to transmit photographic material.

At the receiving end, the signals are received, and the paper is advanced in synchronism with the transmitted document. The image is transferred to the paper by one of several methods, the commonest of which is by thermal means. A special paper, when subjected to heat, turns black. The printing mechanism contains a large number of tiny heating elements which enable the paper to be selectively heated, thus forming the image.

Other printing mechanisms, mostly derived from computer printing devices, enable printing on plain paper, generally providing a cleaner and more durable copy.

Figure 8.3 is a block diagram of a typical fax machine. Note that the transmission and reception of image information is handled by a modem, and the whole operation is supervised by the ubiquitous microcontroller. Since all the component parts are already available, most fax machines may also be used as photocopiers. However, the quality of reproduction is not as good as a dedicated photocopier if the thermal process is used, and the source material must be in single sheet form.

Fax machines have progressed through a number of standards over the years, all the while improving in performance. The standards are recommended by the CCITT (International Telephone and Telegraph Consultative Committee). CCITT Group 1 (1971), which is now obsolete, required some 6 min. to transmit a single page of A4. Group 2 (1976), which is becoming obsolete, reduced this time to about half, while Group 3 (1980) reduces the transmission time still further to under 1 min. Part of this speed increase was made possible by the ability to 'skip over' blank areas of the image, allowing the paper to be advanced more quickly. Group 4 facsimile is a standard for ISDN communication where a page may be transmitted in 3 or 4 s.

Features

Fax machines vary in price considerably, depending upon the features incorporated. Over and above the basic requirements, machines may offer such facilities as a handset, a paper-cutting mechanism, automatic redial if the destination number was busy, automatic transmission of a single document

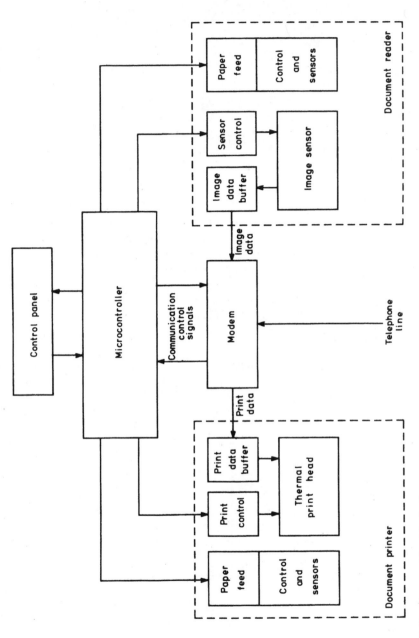

Figure 8.3 *Fax machine block diagram.*

to a large number of destinations (usually for advertising purposes), dialled-number memories of various sizes, and many others.

It should be noted that problems may arise if attempting to connect a fax machine and a TAM (telephone answering machine) to the same line. Both devices are designed to answer incoming calls automatically, and some consideration to the problem will avoid contention. If single-line use is envisaged, choose a fax machine with a TAM interface (sometimes called a TAD (telephone answering device) interface) or one with a built-in TAM. Alternatively, there are a few fax switches on the market that can detect incoming fax tones and route the call to the fax machine, but these tend to be expensive and not entirely satisfactory.

Installation

Since different models of machine offer different facilities, the amount of setting up and programming will vary. In most cases, the setting up will be limited to connecting together the component parts (power supply – if separate – and handset, etc.), installing the paper, and checking the operation. In some cases, where the fax machine is to be used as well as telephones on a single PSTN line, some re-routing of connections may be necessary, if the fax machine includes fax/voice switching (see Fig. 8.4).

Programming will probably involve setting the date and time (which may be required for time-stamping the messages) and setting the user's name and fax number which will automatically be sent, on both transmission and reception, to the distant end. Other setting requirements will be model-dependent and the manufacturer's instructions need to be carefully studied.

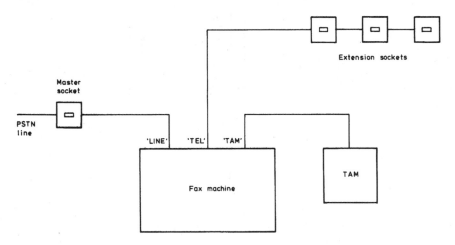

Figure 8.4   *Typical fax machine installation for single-line use.*

Payphones

Introduction

Like all telephone equipment, payphones have changed considerably in their manner of operation over the years. Originally, charging was controlled from the local exchange so that tariff changes could be accommodated more easily. This led to the development of fairly complex signalling between the exchange and the payphone in the form of meter pulses. However, the increased use of privately owned payphones together with sophisticated microprocessor-based technology has led to call charges being controlled at the payphone. Some privately owned payphones still rely on meter pulses for their operation, but these are becoming obsolete, and meter pulses are being phased out.

Choice of equipment

Privately owned payphones are becoming more popular in a number of environments, for a number of reasons. In the simplest case, a low-cost payphone might be installed in a private home to act as a reminder that telephone calls cost money. Such payphones will probably only accept one value of coin and will exhibit a low degree of security.

At the other end of the scale, a sophisticated payphone will accept all current coins of the realm, and be capable of 'learning' new coins that are introduced. They will also be able to charge appropriate rates for any type of call at any time of day. Some have a 'follow-on' button, which allows a further call to be made if there is unused time remaining. By means of programming, the owner will be able to prevent certain types of call, such as premium rate calls to chat lines, from being made.

It is worth noting that very few payphones are suitable for installation in outside locations, including ex-BT telephone kiosks. They are not sealed against moisture, and the internal circuitry is not protected against condensation. Rapid deterioration of the equipment is inevitable if such locations are chosen.

Most payphones, but not all, are suitable for working behind a PABX – that is, they may be programmed to dial automatically the appropriate code to access an outside line. If such usage is intended, this aspect should be checked.

Security

Payphone security may be divided into the following two aspects:

Tampering and vandalism

Where the intention is either to gain unauthorized access to the cash box with theft in mind, or simply to cause damage, there are few payphones that can withstand sustained attacks. There are a few measures that can be taken, however:

- If possible, the payphone should be sited in a supervised location.
- For some models, steel cases with large cash collection boxes are available at extra cost.
- If there is an existing intruder alarm on the premises, one might consider fitting a reed switch and magnet so that if the payphone is opened, the alarm is sounded.

Fraud

To prevent fraudulent use of the telephone line by simply unplugging the payphone and using another telephone in its place, there are several possibilities. The socket may be mounted in an inaccessible position – some payphones are designed to be attached to a wall over the socket for instance – or the socket may be fitted with a lock to prevent removal of the plug (see 'Accessories', later in this chapter).

In some instances, though, it may be desirable to have a portable payphone. Small guest-houses, for example, may provide sockets in guest-rooms with a payphone available on request. In these cases, plugs and sockets may be replaced with the reverse-clip versions (see Chapter 2).

Most payphones will emit 'cuckoo tones' for a few seconds after the handset has been lifted. This is a fraud prevention device. A reverse-charge call will not be connected by an operator if these tones are present.

Legal requirements

A number of legal requirements govern the installation of payphones.

- The owner must display the call charges, and most manufacturers provide a suitable notice to meet this requirement.
- If the payphone is connected in parallel with other phones, the user must be made aware that his or her calls may be overheard.
- Emergency (999/112) calls must be available free of charge. In many cases, a full cash box, or a problem with the coin collection mechanism, results in the display '999 calls only'.
- Payphones must not be connected to the same circuit as certain series-connected devices such as fax machines.

Programming

The installation of a payphone will require a certain amount of programming in accordance with the manufacturer's instructions. Such programming will usually include setting of date and time (for correct call charging), tone or pulse dialling, PABX access code (if required), call-charge rates, and local dialling codes. Note that unless the date and time have been set, as a minimum, the payphone will not operate correctly.

Installation tips

When programming a payphone for PABX operation, the pause between the dialling of the PABX access digit(s) and the dialling of the required number may be adjustable. The default value may be as high as 3 s. Certain overseas calls may fail to be connected unless this pause is reduced to 1 s.

It is difficult to remove coins from older GN Rathdown Solitaire pay-phones – they must be scooped out a few at a time. It has been found that a small margarine tub fits very well into the coin collection space, and the whole tub may be lifted out, emptied, and replaced within a few seconds. Later models are supplied with a coin tray.

Accessories

Audible and visual indicators

There are many occasions when the call arrival indication built in to standard telephones is unsuitable or inadequate. In these circumstances, alternative or additional devices are available.

The BT 50E bell is a twin-gong bell that is intended for interior use only, and is equipped with a flying lead to be plugged into a standard telephone socket. It will primarily be found useful in domestic installations where the householder may not be able to hear the telephone ringing in all parts of the house.

For outside use, the BT 80D bell is suitable, being both weatherproof and somewhat louder. This type is designed to be hard-wired between connections 3 and 5 of a master or secondary socket. Note that the concentric bell gongs are mounted by means of a screw that is slightly off-centre. The gongs are rotationally adjusted during manufacture to optimize performance, and therefore the gong screw should not be disturbed.

There is also a 'tone caller' available. This is a small low-cost device with an integral BT 431A plug and LED, and is simply plugged directly into a standard telephone socket. It produces a similar tone to that of a modern telephone, and it may be found to be useful in some circumstances.

In cases where there is a very high ambient noise, where silence must be maintained, or where the user is hard of hearing, a high-intensity flashing light may be found most useful. One type available is activated by means of a microphone attached to a telephone, and produces flashes, activates a high-power sounder, or both, in sympathy with the ringing tone. The mode of operation is switch-selectable. Another type, which flashes only, plugs directly into a telephone socket.

There is also a range of hooters, klaxons, sirens and flashing beacons available. Not all of these are suitable for direct connection to the PSTN or PABX – some are designed for 12 or 24 V operation, and require additional power supplies and/or ring detectors to enable them to operate.

REN boosters

Ringing problems due to excessive loading on either direct exchange lines or PABX extensions can be overcome by installing a REN booster. This mains-powered device is fitted between the line and the telephone apparatus, and increases the REN value to eight. In the event of a power failure, the line ringing current is directly connected to the telephone apparatus, and thus it may be necessary to remove some of the load temporarily by unplugging some of the devices.

Socket locks and call-barring devices

The locking bar is a cheap and simple device that is available in two sizes to fit the LJU2 and LJU3 size sockets. It is screwed in place using the existing faceplate mounting screws after the telephone plug has been inserted, and thus removal is prevented. It is recommended for use in supervised locations to prevent the telephone from being removed, or to prevent another telephone from being used in its place.

Slightly more sophisticated is the socket lock, which, although attached by the same means, has a key-operated cover that conceals the screw heads. It may be secured either with a plug in place, or without – in which case, the socket may be blanked off to prevent unauthorized use.

Call-barring devices range in complexity, but generally allow any incoming calls, but outgoing 999 calls only. They may be key or PIN-code operated, and may allow selected calls (e.g. local calls) to be made without restriction. Once fitted, they may also control additional extension sockets.

Privacy sockets and adaptors

The privacy socket is a form of the LJU3 socket, from which it is externally indistinguishable, while the privacy adaptor is similar in appearance to a two-way adaptor but with only one outlet. Both forms operate in the same way – users of extensions fitted with the privacy device are prevented from overhearing conversations on other extensions. Such devices find use where a payphone not already equipped with privacy circuitry shares a line with other telephones, or in manager/secretary situations.

Least-cost routing devices

In order to take advantage of lower call costs offered by alternative network operators, customers must decide which calls are to be routed via that operator, and then to dial, or arrange to dial, a complex access code. This procedure may be made transparent by fitting a 'least-cost routing' (LCR) device that routes the call automatically, depending on the number dialled. The device, that is mains-powered, must be fitted between the exchange line and all extensions. Note that when the device is installed, the familiar 'dialling tone' will no longer be heard. One should also be aware that since some PABXs offer automatic LCR, the provision of additional devices will be unnecessary.

Autodiallers

Activated by an external switch, the security autodialler will dial up to six pre-programmed numbers until one is answered, and then relay a recorded message. In this way, friends or neighbours, or factory key-holders, may be notified of an emergency situation, for example intruders or fire. The activation may be made from an existing alarm system.

An enhanced version is available, which is modified to be actuated by a low-power handheld radio transmitter, and may be used to alert friends or neighbours of a home emergency. Such an arrangement finds use if elderly or frail people need assistance but are unable to reach the telephone.

Both devices are mains-powered with a back-up battery supply in case of power failure, and are plugged directly into the telephone line.

Change-over switches

A two-pole change-over switch, BT code number 2A, may be used to switch a line or extension to another destination, or to switch an outside bell off at night. These slide switches are supplied with a back-box and are suitable for surface mounting.

Call management devices

Devices that automatically answer incoming calls with a recorded announcement, put the caller on hold, or route the call to the appropriate department, or which log incoming and/or outgoing calls come under the heading of call management.

Except for the simple call-logger, these are mostly aimed at the corporate user and are priced accordingly. The call-logger, however, may be encountered in holiday accommodation, for instance, where it may be used as an alternative to a payphone. A paper 'till-roll' printout of all outgoing calls with destination, time and call charge provides a basis for customer billing. Unlike the payphone, the client does not need cash handy and may make the call from any available extension, and there is no cash box to be emptied or broken into.

9

Tools, test equipment and materials

Essential tools

With any trade, a reasonable selection of tools, both general purpose and more specialized, is a prerequisite to the satisfactory performance of the task in hand. The installation of telephone wiring and equipment is no exception. Firstly, the tools considered to be essential, i.e. without which the installer will be severely handicapped, will be considered.

Screwdrivers

A good selection of flat-bladed, Phillips[1] headed and Pozidriv[2] screwdrivers will be found to be indispensable. One should not neglect the need for some of the smallest sizes – a flat-bladed jewellers' screwdriver is useful for releasing the latches of 'Western Electric' style connectors, for instance. The following should be considered to be a minimum.

Flat blade:	3 mm, 5 mm, 8 mm
Phillips:	No. 1, No. 2
Pozidriv:	No. 0, No. 1, No. 2
Jewellers':	Set of six.

Pliers and cutters

For general wire-forming and positioning, a pair of snipe-nosed pliers will be found to be most useful, and for rather heavier work, a pair of combination pliers will find many uses.

[1] 'Phillips' is a trademark of the Phillips International Corporation (USA)
[2] 'Pozidriv' is a trademark of European Industrial Services Ltd.

Two pairs of wire cutters are recommended. The first should be a good-quality pair of electronic grade side cutters that are capable of cutting up to three-pair CW1308 cable as well as trimming individual wires. For cutting heavier cables, and particularly for cutting CW1378 dropwire because of the integral steel strainers, a pair of heavy-duty cutters is essential. These should be specifically rated as suitable for 1.5 mm piano wire at least. If standard electronic grade cutters are used to cut steel wire, they will very soon become so damaged as to be unusable.

A pair of wire strippers suitable for use with 0.5-mm-diameter conductors is another essential item. Even though insulation displacement connections require no stripping, there will be occasions where screw terminals need to be used, and where wire ends must be stripped for testing purposes.

Other general tools

The installer's toolbox will contain a sturdy craft knife, a small pair of sharp pointed scissors, a hammer, a tape measure, a junior hacksaw, and a hammer-action power drill with the usual selection of HSS and masonry bits. A number of extra-long bits should be added to this selection, so that holes may be drilled in door and window frames, and through walls and floors. Initially, 300-mm-long wood and masonry bits should be obtained in two sizes – 6.5 and 13 mm diameters.

Specialized tools

Probably the single most important tool for installation work is the insertion tool shown in Fig. 9.1. It is used for making connections to the IDC terminals of sockets and connection boxes, and is frequently referred to as a 'Wire Inserter 2A' using the BT identification code number. The manufacturer of the tool is Krone (UK) Technique Ltd., and their part number is 6417/1/810/02. The tools are available from all of the suppliers listed in Appendix 3, at prices generally in the range of £12–£15. (Prices quoted in this chapter were correct in October 2000.)

There are two features of this tool worthy of mention. Firstly, there is a scissor-action cutter that is designed to cut off surplus wire after the termination has been made. This occurs automatically as part of the punch-down action. It is necessary to take care when terminating wires onto 237-type connection strips, that the small pieces of cut-off ends do not find their way into the test jacks that lie between the two rows of terminals. Occasionally, it is desirable not to cut the wires at all, but to leave them long enough to continue on to another terminal. If this is the case, the cutting action can be prevented by attaching a small plastic clip around the metal portion of the tool at the point where it emerges from the plastic handle. This clip is supplied with the tool, attached to it by means of a short length of nylon cord.

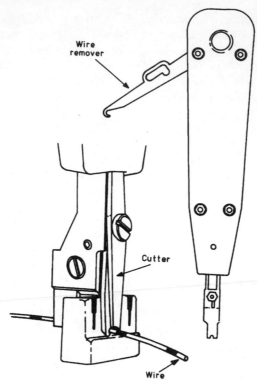

Figure 9.1   *Professional IDC insertion tool.*

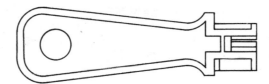

Figure 9.2   *Disposable IDC insertion tool.*

Secondly, concealed within the handle of the tool is a fold-out metal hook for removing wires from terminals. The wire to be removed should be held under some tension before extraction, and the terminal should be inspected, and cleared if necessary, if any of the wire or insulation is left behind.

Disposable plastic insertion tools (see Fig. 9.2) have a limited life, but it is always worth having one or two in the toolbox for emergency use. Note that they are intended to be used one particular way round, as the IDC terminals are not centrally located in the plastic moulding. The correct orientation is best determined by inspection – the 'heel' of the tool should be on the same side of the terminal as the wire entry.

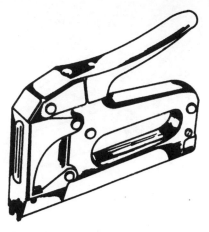

Figure 9.3 *Cable tacker.*

From the most important tool to the cheapest. A short length of galvanized wire about 1 m long, as used for the erection of wire fences, will be found most useful for threading cables through holes in walls, floors and ceilings. The end of the wire is fed through the hole and forced into the end of the cable. After binding the joint with plastic insulating tape, the cable can be pulled through the hole usually without much difficulty. The same piece of galvanized wire, with a hook formed at one end and perhaps lengthened with a stick or piece of trunking lid, can be used to retrieve cables across roof spaces and other inaccessible locations. It can also be used as a temporary lashing during the installation of overhead cables, clearing a way through brick and plaster rubble, and many other purposes. Don't leave home without it!

The neatest and quickest method of attaching cables to skirting boards, plasterwork, and so on is to use a staple gun or 'tacker'. An example is shown in Fig. 9.3. These are readily available from numerous suppliers at prices ranging from about £25 to £45. Models are available for different sizes of staples – the 4.5 mm size is the most useful as it is suitable for use with three-pair CW1308 cable. Staples may be obtained with either a zinc–tin alloy coating or a white coating. The latter are ideal for indoor use and render the staples almost invisible. Note that this method of cable attachment is not suitable for mains wiring, and it should never be used if there is the slightest risk of penetrating cables carrying dangerous voltages. For safety reasons, it is recommended that eye protection be worn when using staple guns.

Crimp tools for assembling BT431A and 631A plugs may be obtained from any of the suppliers listed in Appendix 3, and one should expect to pay between £30 and £50. Some types have integral strippers for removing the outer sheath, while others have guide markings to aid preparation.

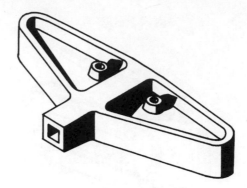

Figure 9.4   *Dropwire No. 10 sheath stripping tool.*

For about £15, a tool may be purchased for the removal of the outer sheath of CW1378 (dropwire No. 10). The tool, shown in Fig. 9.4, is pushed over the cable and then pulled back. An internal blade splits the sheath for the required length so that the conductors can be extracted, and the sheath and steel support wires removed.

Tools for safety and access

Depending upon the environment in which one is working, a safety helmet (hard hat) may not be just a 'good idea', it may be mandatory. A hard hat should of course be worn whenever there are likely to be others working above. Ensure that the helmet is manufactured to BSEN 397, and expect to pay around £6 or £7.

For about the same figure or even less, a pair of safety spectacles or goggles may be obtained for protection when hammering masonry nails (as used in cable clips) and when using staple guns or drilling. Safety spectacles and goggles should conform to BS2092.

If any pole work is anticipated, a pole belt must be used – again, in the interests of safety. This consists of a strong webbing belt that fits around the waist. A second strap, attached to the belt on one side, is passed around the pole when the working height has been reached, and clipped onto a loop on the other side of the belt. The slack is taken up using a releasable ratchet arrangement, and then the wearer is free to use both hands while leaning back against the belt. These rather specialized items can be obtained from Comtec – the address is given in Appendix 3.

A step ladder and an extending ladder will both be needed frequently. As safety is an important consideration when working at heights, look out for ladders conforming to BS2037.

For relatively long-term illumination in difficult conditions, an inspection lamp with an extension lead will prove invaluable, while for the shorter term, a battery-powered torch is indispensable. In particular, the pocket-sized 'Mag-Lite'[3] with its variable-focus beam and robust construction is ideal for the purpose and will cost around £13–£14.

Another useful source of illumination is the head torch, usually powered by a rechargeable battery pack, and attached to the head or safety helmet by means of straps. This is probably the most effective way of illuminating junction boxes while keeping both hands free. Prices start at around £20.

Desirable tools

Screwdrivers

An appreciable amount of installation work inevitably involves the use of screws, and in most cases, the job will be made quicker and easier if a cordless screwdriver is available. It will be found that the screws supplied to secure telephone socket faceplates to the back boxes are both long and fine-threaded, and therefore rather tedious to remove and replace. With this in mind, it is as well to ensure that the screwdriver bits are chosen to include a 4 mm flat blade as well as the more usual sizes and types.

Pliers

The successful and consistent crimping of PJ-filled splices is dependent upon even pressure being applied to the capsule. While crimping is possible with a pair of ordinary combination pliers, the job is made easier if a pair of cantilever action parallel-jawed pliers is used. This should prevent the plastic insert from being tilted to one side and allow the terminals to engage the wires fully.

Other general tools

Tool belts are used in many trades, especially where tasks need to be performed in various locations. They consist of a belt, provided with various pouches and loops, and allow a selection of hand tools and small components to be carried easily, leaving both hands free. The use of a tool belt is particularly recommended when working at height, and especially for pole work. Price: around £20.

Specialized tools

Attention has already been drawn to the need for clear, accurate, and durable labelling, on sockets, connection boxes, and cables. There is now a wide range of electronic labelling devices, powered by rechargeable batteries, which produce pleasing and professional-looking results. The 'office

---

[3] 'Mag-Lite' is a trademark of Mag Instruments Inc.

equipment' section of both the RS and Farnell catalogues (suppliers' addresses in Appendix 3) is the place to look, where prices start at about £40.

A cable dispenser is a stand for holding a drum of cable, allowing the cable to be unwound easily, without kinks or tangles. These are available from a number of sources or alternatively, quite easily made from wood or rigid plastic waterpipe components.

A special assembly tool is made by from Krone UK (Technique) Ltd., and available from several of the sources in Appendix 3, to ease the wiring of type 237 connection strips. The 10-pair 'outrigger', as it is termed, has the Krone part number 6517/1/842/00, and an eight-pair version is also available, part number 6417/843/00. The outrigger is used to position a connection strip some distance away from the backmount frame, allowing the cabling to be positioned more easily and with a uniform amount of spare cable. Typical pricing is around £11.

Essential test equipment

The multi-function test meter, or multi-meter, is a most basic and indispensable piece of equipment. It will be required for DC voltage measurement (up to 100 V), AC voltage measurement (up to 250 V) and resistance/continuity testing. The degree of accuracy required is not great and will be easily exceeded by any meter on the market. More important will be the ability of the instrument to withstand the rigours of the toolbox and of the open air, and for this reason, the digital instrument (which in any case has largely superseded the analogue moving-coil meter), will be the more robust. It is a matter of individual choice whether to purchase the cheapest possible at, say, £15 or so and be prepared to replace it frequently, or to spend £200 or more on a robust instrument and hope that it does not get lost or stolen.

Some method of testing telephone sockets will be needed. There are a number of simple testers that indicate the presence of line voltage, and its polarity, by means of LEDs. These do not test for the presence of the bell-wire connection to terminal 3 and may not operate satisfactorily with the lower-voltage characteristic of smaller PABXs and, for these reasons, are less than satisfactory. For those readers with some electronic expertise, details of a socket tester that overcomes these problems are given in Appendix 4. Strictly speaking, no device of the 'home-made' variety may legally be connected directly or indirectly to the PSTN. The circuit diagram is therefore only offered for information purposes.

Most installers will need to equip themselves with a test telephone, or 'butt' (the term taken from the ability of the user to 'butt-in' to a line) (see Fig. 9.5). As the basic functions of a telephone (ring, dial, talk, and listen) are incorporated, as well as line polarity testing, this piece of equipment will find frequent application. Several different models are available

Figure 9.5   *Test telephone 284/2.*

Table 9.1   *Comparison of test telephone features*

	284/2	DSTS-2	UTS-E	UTS-ELS	UTS-3
Dialling: LD	Yes	Yes	Yes	Yes	Yes
MF	Yes	Yes	Yes	Yes	Yes
Redial	No	Yes	Yes	Yes	Yes
Recall type	ELR/TBR	TBR	TBR	TBR	TBR
Memories	0	10	10	10	10
Polarity test	Yes	Yes	Yes	Yes	Yes
Ringer	No	Yes	Yes	Yes	Yes
Monitor	Yes	Yes	Yes	Yes	Yes
Amplified loudspeaker	No	No	No	Yes	Yes
Hands-free operation	No	No	No	No	Yes
Internal battery	No	No	Yes	Yes	Yes
Impact resistant	No	Yes	Yes; 3 m drop onto concrete		
Sealed to IP64	No	No	Yes	Yes	Yes
Price guide	£100	£100	£125	£135	£195

from the suppliers in Appendix 3, and a selection is shown in Table 9.1 for comparison purposes. One or two of the features of these specialized telephones require some explanation.

Firstly, all models offer a 'monitor' facility, whereby the activity of the line being tested may be monitored in the earpiece or loudspeaker without affecting the line conditions. In the monitor mode of operation, there is no local loop current flowing through the test telephone, so it is considered to be 'on hook'. It is not possible to dial, and the microphone is inactive. It is a useful mode for identifying idle and busy lines without disturbing callers.

Secondly, the 284/2 does not have an integral ringer, as in the other models. By switching to the monitor mode, the ringing signal becomes audible, and the polarity indication LEDs will flash. This is satisfactory when testing for the presence of a ringing signal, but limits its usefulness when it is necessary to use it as an ordinary telephone – the sound level is simply not high enough.

Thirdly, there are variations between models in the environmental specifications. All of the units are robust, when compared to ordinary subscribers' equipment, as might be expected. It is claimed that the UTS series of test telephones will withstand a 3 m drop onto a concrete floor, the DSTS-2 is claimed to have a 'high impact case', while the 284/2 is merely referred to as 'tough'. There are similar differences in the degree of sealing against dust and moisture. The UTS series is sealed to IP64 (a recognized standard indicating total protection against dust, and that the unit will withstand low pressure water jets from any direction). The DSTS-2 is 'water-resistant', and no claims at all are made in respect of the 284/2.

All of these test telephones are fitted with a line cord and BT431A plug as standard. The need will arise from time to time to be able to connect the test telephone either directly to stripped wire ends, or to test jacks on an NTTA or MDF. A number of adaptor leads may therefore need to be made up. For connecting directly to stripped wire ends, connect two flying leads from terminals 2 and 5 of a line socket and terminate the other ends of the leads with crocodile clips. For accessing the test jacks of an NTTA (241 strips), connect a line socket as before to the flying lead of a Krone D95737 test cord; and for the 237-type strips, leads made up with D94469 (two-pole) and D94471 (four-pole) test cords will be needed.

The two-pole plugs enable circuits to be monitored without breaking the connection, while the four-pole plugs, when used with disconnection strips (type 237A), will break the circuit under test and simultaneously allow connection to either or both sides of the circuit so that cable and other faults may be isolated. All of these adaptor leads are available, ready made, from Comtec – the address is given in Appendix 3.

Desirable test equipment

Sooner or later, the need will arise to be able to locate and identify cables, and wires within cables. This often occurs because of inadequate record-keeping and the problem of cables becoming inaccessible after installation. There are four different types of cable testing equipment.

The cable locator is essentially a compact metal detector that will locate concealed metallic objects. These detectors can be purchased from hardware stores, and their primary purpose is in locating mains cables and pipes prior to nailing or screwing into walls. For a model aimed at the DIY market, prices will be around £15. For a professional unit, with higher sensitivity and selectivity, one may expect to pay about £70.

The cable tracer is a two-part device. A tone transmitter is connected to one end of the cable, and a hand-held probe is used to detect the tone along the whole length of the cable, assuming that there are no breaks. The probe only needs to be in close proximity to the cable, and the audio output changes in volume according to the distance from the cable. It is sufficiently discriminating to enable one pair from a bundle of pairs to be isolated. For a good-quality instrument in this class, one must expect to pay around £200.

Individual wires in a multi-pair cable can be positively identified using a pair identifier. Again, this is a two-part device. At the distant end, the stripped ends of each wire are attached individually to a small unit by means of numbered and colour-coded crocodile clip leads. Typically, up to 19 wires may be attached, plus a common return. At the near end, the second unit is connected, and each wire is probed in turn. The number displayed on an LCD corresponds to the number of the crocodile clip to which the wire is attached at the far end. It is a particularly useful device when the pairing of CW1296 cable has been lost, or where there may be concealed joints with wire-colour changes along the route. These items of test equipment are priced at about £70.

Cable fault locators are expensive items, with models available between £800 and £2000. The principle of operation is to transmit a pulse from one end of the cable and to measure the time taken for an echo to be received from a discontinuity – a short or open circuit. The technique is known as time-domain reflectometry (TDR), and its primary use is in locating faults in underground cables, especially where access to the ground above the buried cable may be difficult or dangerous. Obviously, the time and labour saved in being able to locate a fault to within a metre or so will justify the cost of purchase, but only if it is felt that some work is likely on long underground cables. Even though a fault locator of this type can also be used to measure the remaining length of cable on a partly used drum, the equipment is probably of greater interest to PTOs than to installers.

For easy electrical access to cables that are already terminated on telephone sockets, simple breakout adaptors can be purchased for just a few pounds. These are used to extend all six socket connections either to 4 mm test lead sockets, or to six phosphor–bronze pads housed in a plastic moulding to which test probes or crocodile clips may easily be connected (see Fig. 9.6).

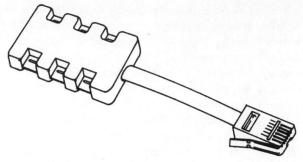

Figure 9.6   *Line socket breakout adaptor.*

Where PABX installation is concerned, the integrity of the mains socket earth should be tested as a matter of course. For this test, an earth loop tester will be required, preferably of the variety that does not trip residual current circuit breakers. The maximum value of earth loop resistance should not exceed 2 Ω, or the value laid down in the current IEEE Wiring Regulations for the particular type of electrical installation. Typically, the price will be in the range of £200–£300.

When two people are working together, tackling problems of installation and fault-finding, a pair of two-way radios ('walkie-talkies') can be invaluable. These are preferred over the alternative linesman's telephones since they do not need to be set up, there does not need to be a serviceable cable in place, and the users are able to move around more freely. Cheaper models are to be avoided since performance is likely to be poor. The Motorola 'TalkAbout'[4] models work well up to a range of about half a mile, and are licence-exempt. The cost of a pair of these units is about £170 + VAT.

Materials

Anyone aspiring to offer a telephone installation service will need to be equipped with some basic materials, although if large quantities, or specific items not listed here, are required for particular jobs, they can usually be obtained within 24 h from the specialist suppliers.

Wire and cable

At the top of the list is a 200 m reel of white CW1308 three-pair cable that will be used in alarming quantities for all types of installation. The same quantity of two-pair 'dropwire No. 10', for external and suspended use should also be available, as should a 100 m reel of cream 1.5 mm earth

[4] TalkAbout is a trademark of Motorola.

wire. These three types should suffice initially, and then as the need arises, larger cable sizes (six-pair, 10-pair, etc.) can be added. Flat under-carpet cable (CW1316) is quite expensive (about seven times the price of the CW1308 equivalent) and so its purchase can be deferred until it is actually needed. Twisted-pair jumper wire is useful if one is wiring larger systems, but for the occasional modification, a pair can be easily extracted from an offcut of CW1308.

Terminations

Sockets

A good selection of sockets should be available. The LJU2/3A will be needed more than others for simple extensions, and the LJU2/2A for PABX extensions. Smaller quantities of the larger LJU3, LJU4, and LJU5 sockets may be needed to match existing sockets, to make use of existing back-boxes, or to provide dual outlets. Back-boxes may need to be purchased separately, especially with the larger socket sizes.

Block terminals

A few block terminals – type 77A (three-pair) and 78A (four-pair) should be stocked, and a quantity of two-wire and three-wire PJ-filled splice connections should be available. Occasionally, and especially for joins of a temporary nature, screw terminal blocks ('chocolate blocks') will be useful, although these should not be left in service for any length of time as the electrical connection is not weatherproof or gas-tight, and the terminals may rust.

For external use, one or two 66B block terminals should be kept to hand.

Fixings

For securing LJU back-boxes and connection boxes in place, a good selection of woodscrews, together with wall plugs and cavity fixings, must be available. Twin-threaded, Supadriv[5] headed screws make life easier, in a selection of lengths from $\frac{1}{2}$ inch to $1\frac{1}{2}$ inch; no. 6 and no. 8 are recommended as a minimum.

Nylon cable ties, particularly the smaller sizes, are constantly used, and a selection of cable clips will be needed. For three-pair cable, use 4 mm clips – white for indoor use and black for two-pair dropwire. For six-pair cable, use 6 mm clips, and for 10-pair cable, the 8 mm size will be found to be satisfactory.

It is as well to carry a few lengths of the smaller sizes of white PVC trunking, with a selection of angles, tees, and end stops. Apart from helping to keep an installation neat, the use of trunking can overcome problems of

[5] Supadriv is a trademark of European Industrial Services Ltd.

proximity to mains cabling. Self-adhesive trunking is slightly more expensive than plain trunking, but installation time will be saved as the drilling, plugging, and screwing will not be required.

Finally, a couple of rolls of self-adhesive PVC insulating tape will always be useful. White tape can be written on with a ballpoint pen, and thus temporary identification labels can be made up quickly and cheaply.

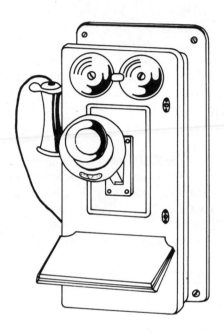

'Magneto Wall Set' U.S.A. 1907

Appendix 1
Glossary

Abbreviated dialling See **Speed dialling**.

AC Alternating current.

Activity report FAX term. A printout of a list of incoming and outgoing messages with length and time of transmission. May be produced automatically or on demand.

AF See **Audio frequency**.

Announce only TAM mode for use on information lines where no incoming message is to be recorded.

Audio frequency (AF) A range of frequencies discernible by the human ear, approximately 20 Hz to 15 kHz.

BABT British Approvals Board for Telecommunications.

Bandwidth A range of frequencies that a circuit will pass.

Battery back-up Prevents loss of memory contents in the event of power failure.

BORSCHT An acronym representing the essential features of a telephone exchange: Battery, Overvoltage, Ringing, Supervision, Coding, Hybrid and Test.

BRA Basic rate access (ISDN).

BRI Basic rate interface (ISDN).

Broker's call Feature of a PABX allowing a user to make a call to an internal extension during an outside call, and to switch between the two while maintaining privacy.

BS British Standard.

BSI British Standards Institute.

BT British Telecom.

Butt A telephone with various connection adaptors for testing purposes.

Cable An assembly of two or more conductors insulated from each other and enclosed in a protective sheath.

Cadence A ringing pattern.

CAI Common air interface (also **Call arrival indication** q.v.).

Call arrival indication Ringing signal sent to the called terminal.

Call back when free See **Camp on busy**.

Call barring Prevents outside calls being made or limits types of calls that may be made.

Call diversion PABX feature that allows calls to be automatically re-routed from one extension to another either immediately or if the call is not answered within a set number of rings. Also called 'Follow-me' or 'Call forwarding'.

Call forwarding See **Call diversion**.

Call intrusion PABX feature allowing the user to break into another call.

Call logging See **Call management**.

Call management PABX feature to allow all incoming and outgoing calls to be logged for control or charging purposes, or for traffic flow analysis.

Call monitor See **Call screening**.

Call pick-up PABX feature allowing a user to answer a call ringing on another extension.

Call screening TAM mode where incoming calls can be heard over the loudspeaker, and the user may decide whether to answer the call or not. Also called **Call monitor**.

Call splitting Same as **Broker's call**.

Call transfer PABX term. The re-routing of a telephone call on an outside line to a different extension.

Call waiting PABX or PSTN feature whereby a user is notified during one call of another call requiring attention, usually by means of tones.

Calling line identification (CLI) Network facility where a caller's number is passed on to the recipient.

Camp on busy PABX feature. If an internal call fails because the called extension is busy, setting this feature will initiate a ring back and automatic re-dial, once the called extension handset is replaced. Can also be used during busy periods to queue for an outgoing exchange line.

Catenary wire A support wire for overhead cables, not used electrically.

CCIR International Radio Consultative Committee.

CCITT International Telegraph and Telephone Consultative Committee.

CCU Central control unit.

Central office (CO) US term for the local telephone exchange.

Centrex PABX facilities provided by the local exchange.

Centronics Parallel computer interface commonly used to connect to printers.

CEPT European Conference of Postal and Telecommunications Administrations.

CLI See **Calling line identification**.

CO See **Central office**.

Code, security Electronic 'handshake' between handset and base unit of a cordless telephone.

Compander (Compressor and expander) A method of audio noise reduction used in cordless telephones.

Conference call Several parties connected together in one telephone conversation.

CPE Customer premise equipment.

Crossbar switching A method of interconnecting signal paths by means of a horizontal/vertical matrix of switch contacts.

Crosstalk Unwanted signals passing from one circuit to another.

CT Cordless telephone.

CTI Computer–telephony integration.

CT1 Cordless telephone standard.

CT2 Cordless telephone standard.

DASS Digital access signalling system.

Day service PABX feature. One of two modes of operation, usually equating to attended service, to allow one set of PABX programming options to be used. Night service provides a different set of options for ringing assignments, etc.

DC Direct current.

DCE Data Communications Equipment.

DDI See **Direct Dialling In**.

DECT Digital European cordless telephone.

DEL See **Direct exchange line**.

Digital OGM TAM feature where the outgoing message (OGM) is recorded in a memory chip rather than on tape. The caller has less time to wait as the tape does not have to rewind, and the message does not deteriorate in time.

Direct dialling in (DDI) Allows external callers into a PABX to dial directly to the required extension without going through a switchboard operator.

Direct exchange line (DEL) An outside line connecting to the PSTN.

Direct inward system access (DISA) PABX feature allowing an outside caller to access system features and extensions.

Direct station select (DSS) PABX hardware feature whereby the user may dial extensions with a single button push.

DISA See **Direct inward system access**.

DND See **Do not disturb**.

Do not disturb (DND) PABX feature allowing a user to prevent incoming calls.

Door opener PABX facility allowing a caller access to the building by electrically unlocking the door. Arranged through the PABX by dialling on an extension telephone.

Door phone PABX facility allowing communication from a PABX extension to a person at the door.

DP Distribution point.

DPNSS Digital private network signalling system.

Dropwire External grade cable suitable for overhead suspended use with no additional support.

DSS See **Direct station select**.

DTE Data terminal equipment.

DTI Department of Trade and Industry.

DTMF See **Tone dialling**.

Dual channel operation Allows change of cordless telephone radio channel in the event of interference.

Dual tape TAM feature where there are separate tapes for the OGM and the ICM. The caller has less time to wait as the tape does not need to be repositioned.

Duplex See **Half duplex** and **Full duplex**.

Earth loop recall (ELR) One of two types of recall signal to the PABX from an extension when hold or transfer is required. Operates by connecting the 'a-wire' to earth.

Electronic switching system (ESS) A telephone circuit switching device controlled electronically (PABX).

ESS See **Electronic switching system**.

ETACS Extended total access communications system (mobile-phone standard).

Eth Abbreviation for telecomm. functional earth.

ETSI European Telecommunications Standards Institute.

Fax Facsimile.

Fax/Tel switch Fax term. A switch to allow either fax operation or voice operation.

FCC Federal Communications Commission.

FDM See **Frequency division multiplex**.

FE Functional Earth.

Flash hook A button, or action, to disconnect the local loop connection temporarily. Sometimes used to signal to the exchange to put the caller on hold.

Hands-free dialling Same as **On-hook dialling**.

Follow me See **Call diversion**.

Follow-on call Payphone term. On completion of one call, pressing this button allows a second call to be made with the unused money already inserted.

Frequency division multiplex (FDM) A method of combining two or more communication channels by frequency-sharing technique.

Frequency shift keying (FSK) A method of data transmission commonly used in modems where the two states of the signal are transmitted as tones of two different frequencies.

FSK See **Frequency shift keying**.

Full duplex A circuit that carries information in both directions simultaneously.

Group working PABX feature allowing groups of extensions to be formed for routing of incoming and outgoing calls. Hunt groups allow calls to be evenly distributed over a number of extensions.

Half duplex A circuit which carries information in both directions, but not simultaneously.

Hands-free operation Numbers may be dialled and conversations take place without the need to lift the telephone handset.

Hearing-aid compatible Compatible with hearing aids that can be switched to the 'T' position to improve speech clarity and eliminate background noise. The coupling is inductive (i.e. magnetic) rather than acoustic.

Highway, Business/Home BT proprietory implementation of ISDN 2e for small businesses/domestic customers.

Hold A method of disconnecting the speech path between callers without breaking the local loop.

Hook switch A switch operated by the action of replacing a telephone handset in its cradle, or by the electronic equivalent.

Host/subsidiary working See **Piggy-backing**.

Hunt groups See **Group working**.

Hybrid operation PABX facility allowing both two-wire telephones and key telephones to be used.

Hz Hertz (= cycles per second).

ICM See **Incoming message**.

IDC Insulation displacement connector.

Incoming message (ICM) Message left by a caller on a TAM.

Inductive coupler Part of a telephone that provides hearing-aid compatibility (q.v.).

Intercom Allows conversations between the handset and base unit of a cordless telephone. Also applied to internal calls between extensions of a PABX.

Internal calls See **Intercom**.

ISDN Integrated services digital network.

ISO International Standards Organization.

ISP Internet service provider.

ISPBX Integrated services private branch exchange.

Keystation See **Keysystem**.

Keysystem A telephone system that uses only special, proprietary telephones, or 'keystations'.

Key telephone Same as **Keystation**.

KTS Key telephone system (see **Keysystem**.)

Last number redial (LNR) Telephone feature allowing the last number dialled to be re-dialled by pressing a single button.

LCD Liquid crystal display.

LCR See **Least-cost routing**.

LD See **Pulse dialling**.

Least-cost routing (LCR) Method of determining the cheapest method of placing a call, automatically. Used when accounts are held with more than one carrier.

LED Light-emitting diode.

Line box A trade-name of BT for the NTE5 master socket.

LJU Line junction unit.

LNR See **Last number redial**.

Local loop The two-wire connection between a subscriber and the local exchange.

Loop disconnect (LD) See **Pulse dialling**.

LV Low voltage.

Manager/secretary operation PABX facility where calls for the manager are routed to the secretary, who can then easily transfer the call if required.

MDF Main distribution frame.

Memory A system whereby telephone numbers can be retained for frequent use.

MF See **Tone dialling**.

Mixed mode dialling See **Tone switching**.

MOH See **Music-on-hold**.

MSN Multiple subscriber numbering. Similar to **DDI** (q.v.); applied to ISDN 2e.

Music-on-hold A method by which external callers to a system hear music, tones or messages while waiting to be connected to the appropriate person or department.

Mute See **Secrecy**.

MUX Multiplexer.

Network terminator (NT) Device that terminates the network connection and converts to the S-bus standard (ISDN).

NEXT Near-end crosstalk.

Night service See **Day service**.

NT See **Network terminator**.

NTE Network terminating equipment.

NTP Network termination point.

NTTA Network termination and test apparatus.

NTTP Same as NTTA (erroneous).

Off-hook The condition that nowadays indicates the active state of the local loop.

Oftel Office of Telecommunications.

OGM, digital See **Digital OGM**.

OHD See **On-hook dialling**.

On-hook The condition that indicates the inactive state of the local loop.

On-hook dialling (OHD) Allows numbers to be dialled without lifting the handset.

Outgoing message (OGM) Message played to caller on a TAM.

PA Public address.

PABX Private automatic branch exchange.

Paging (1) Allows alert signal to be transmitted from base to handset of a cordless telephone (but not speech). (2) Allows general announcement to be made from any extension of a PABX. (3) On-site paging uses a radio transmitter so that holders of appropriate receivers can be alerted anywhere within range of the transmitter – within a building or small complex, for instance. (4) National paging relies on radio network so that holders of appropriate receivers can be alerted anywhere in the country.

Pause Inserted in stored number dialling sequence to allow time for switching to take place, especially between the outside line access code and the number required when dialling through a PABX.

PBX Private branch exchange. Often used interchangeably with PABX.

PCB Printed circuit board.

PCI Pre-connection inspection (now obsolete).

PCM Pulse code modulation.

PE Protective earth.

Piggy-back operation PABX term describing the connection of a smaller PABX to a larger.

PIN Personal identification number.

PJ Petroleum jelly.

PMR Private mobile radio.

Port PABX term referring to the system connection point for an exchange line or extension.

POT Plain ordinary telephone (or plain old telephone).

Power fail telephone A telephone that will be automatically connected to a PSTN line when the PABX becomes inoperative due to a power failure.

PPL Phonographic Performance Ltd.

PRA Primary rate access (ISDN).

PRI Primary rate interface (ISDN).

PRS Performing Right Society.

PSK Phase-shift keying.

PSN Public switched network.

PSTN Public switched telephone network.

PTO Public telecommunications operator.

Pulse dialling A system, now obsolescent, of signalling to the exchange the required number by transmitting a sequence of pulses. Also known as loop disconnect (LD) dialling.

RA Radiocommunication Agency.

Remote access TAM feature whereby messages left can be replayed over the telephone line to the user, who may signal to the TAM using special codes.

REN See **Ringer equivalence number**.

Ringer equivalence number (REN) A method of specifying the loading on an exchange line in respect of a ringing signal. A PSTN line will normally support four standard ringers and is therefore said to have a REN of four.

RS232C Serial computer interface commonly used to connect to printers.

Secrecy A push-button on a telephone that disconnects the microphone temporarily, to allow the user to hold a conversation without the other party overhearing. Also called **'Mute'**.

Sidetone The portion of a user's voice that is intentionally sent back to the earpiece.

Simplex A circuit that carries information in only one direction.

Soft keys Buttons on telephone apparatus which can be programmed to perform specific functions.

Speed dialling A feature found both on PABXs and other equipment allowing users to dial long or frequently used numbers by pressing just two or three buttons. Also called **Abbreviated dialling**.

STP Shielded twisted pair. A type of cable used for data network installations.

Subscriber A telephone customer (sometimes abbreviated to 'sub').

Subscriber loop The local loop.

Surge arrestor See **Surge protector**.

Surge protector A device connected to a telephone line to prevent the voltage across the line, or between each wire of the line and earth, from rising to a harmful level. Also called **Surge arrestor**).

SWG Standard wire gauge.

System X A type of digital telephone exchange in common use by BT as part of the PSTN.

TA See **Terminal adaptor**.

TACS Total access communication system (mobile phone standard).

TAD See **TAM**.

TAD interface See **TAM interface**.

TAM Telephone answering machine, also known as a telephone answering device (TAD).

TAM interface Fax feature allowing an answering machine to be connected to the same line as a fax machine, and having the ability to discriminate between the two types of call.

TBR See **Time break recall**.

TDR Time domain reflectometer (or reflectometry).

TE See **Terminal equipment**.

Terminal adaptor (TA) Device to convert between S-bus and another user standard. (ISDN)

Terminal equipment (TE) Equipment such as a telephone or computer that is connected directly or indirectly to an ISDN line.

Tie line A dedicated private wire between two sites of the same organization allowing one site to access the PABX at the other.

Time break recall (TBR) One of two types of recall signal to the PABX from an extension when hold or transfer is required. Operates by breaking the local loop for a specific period of time.

Time division multiplexing (TDM) A method of combining two or more communication channels by a time-sharing technique.

Time stamp TAM feature where each ICM is annotated with time and date of receipt.

TMA Telecommunications Managers Association.

Toll saver TAM feature. During remote access, the owner may determine by the number of rings if there are any messages waiting. If not, he may hang up before the call is answered without incurring charges.

Tone dialling A system of signalling to the exchange the required number by transmitting a unique pair of tones to represent each digit.

Tone switching Allows users connected to pulse dialling exchanges to switch to tone dialling after the connection has been made so as to access tone-mode services. Also called **Mixed mode dialling**.

Touch tone A trademark of BT in the UK for DTMF dialling.

Two-way recording TAM feature. Allows both sides of a conversation to be recorded on the tape.

UTP Unshielded twisted pair. A type of cable used for data network installations.

VF See **Voice frequency**.

Voice channel A communication channel having a bandwidth of 300 Hz to 3 kHz.

Voice frequency (VF) Frequencies in the range of 300 Hz to 3 kHz.

Vox Voice activation. TAMs so equipped will cease recording when speech ceases.

Appendix 2
Further reading

1. British Standards Institute, (current edition). *BS6701*. BSI
2. Bigelow, S. J. (1991) *Understanding Telephone Electronics*, 3rd edn. Sams.
3. Office of Telecommunications (current edition). *The Wiring Code Pt 1 and Pt 2*. Oftel.
4. Texas Instruments (1994) *Understanding Data Communications*, 4th edn. Sams.
5. Winder, S. (1998). *Telecommunications Pocket Book*. Butterworth-Heinemann.
6. Lee, R. (1996) *The ISDN Consultant*. Prentice Hall (NB Some of the contents of this US publication are not appropriate to the UK).
7. Griffiths, J. (1994) *ISDN Explained*. 3rd edn. John Wiley (written by BT staff; quite technical).

Appendix 3
Useful addresses

Organizations and official bodies

British Approvals Board for Telecommunications (BABT)
Claremont House
34 Molesly Road
Hersham
Walton on Thames
Surrey KT12 4RQ
Tel: 01932 222 289
Fax: 01932 251 201
Website: www.babtps.com

British Standards Institute (BSI)
Maylands Avenue
Hemel Hempstead HP2 4SQ
Tel: 01442 230442
Fax: 01442 231442
Website: www.bsi-global.com

Office of Telecommunications (Oftel)
Export House
50 Ludgate Hill
London EC4M 7JJ
Tel: 0207 634 8700
Fax: 0207 634 8943
Website: www.oftel.gov.uk

Phonographic Performance Ltd. (PPL)
Ganton House
14–22 Ganton Street
London W1V 1LB
Tel: 0207 534 1000
Fax: 0207 534 1111
Website: www.ppluk.com

Performing Right Society (PRS)
29–33 Berners Street
London W1P 4AA
Tel: 0207 580 5544
Fax: 0207 306 4455
Website: www.prs.co.uk
(regional offices in: Edinburgh Tel: 0845 309 3090; Peterborough Tel: 0845
309 3090; Warrington Tel: 0870 901 0101)

Radiocommunication Agency (RA)
Wyndham House
189 Marsh Wall
London E14 9SX
Tel: 0207 211 0211
Fax: 0207 211 0507
Website: www.radio.gov.uk

Trade suppliers

Comtec
Cedar Works
Hillbottom Road
High Wycombe
Bucks HP12 4HJ
Tel: 01494 450921
Fax: 01494 464775
email: sales@comtec-comms.com
Website: www.comtec-comms.com
(specialized telecommunication product stockist)

Farnell Components
Canal Road
Leeds LS12 2TU
Tel: 0113 263 6311 (sales)
Fax: 0113 263 3411 (sales)
Website: www.farnell.com
(exhaustive range of general electronic equipment and components)

Krone (UK) Technique Ltd.
Runnings Road
Kingsditch Trading Estate
Cheltenham
Glos. GL51 9NQ
Tel: 01242 264400
Fax: 01242 264488
Tel: 01242 264444 ('Hotline')
Website: www.krone.co.uk
(manufacturers of IDC connection products, and housing and tooling)

Maplin Electronics
Freepost smu 94
PO Box 777
Rayleigh
Essex SS6 8LU
Tel: 0870 264 6000 (sales)
Fax: 0870 264 6001 (sales)
Website: www.maplin.co.uk
(good range of general electronic equipment and components – mostly supplying the home enthusiast)

Nimans Ltd.
500 Broadway
Salford Quays
Greater Manchester M5 2UE
Tel: 0870 444 3100
Fax: 0870 444 3105
Website: www.nimans.net
(specialized telecommunication product stockist)

The Rocom Group Ltd.
Thorp Arch
Wetherby
W. Yorkshire LS23 7BJ
Tel: 01937 847777
Fax: 01937 847788
Website: www.rocom.co.uk
(specialized telecommunication product stockist)

R.S. Components Ltd.
PO Box 99
Corby
Northants NN17 9RS
Tel: 01536 201201 (sales)
Fax: 01536 201501 (sales)
Website: rswww.com
(exhaustive range of general electronic equipment and components)

Appendix 4
Useful circuits

This appendix contains the circuit diagrams, together with descriptive text, of a few telephone-related devices that may be of interest.

Note that it is an offence to connect any unapproved apparatus either directly or indirectly to the PSTN, and that action may be taken against anyone who does so. None of the following circuits is 'approved'.

The circuits are not intended to be constructed by an electronics novice, and so no detailed constructional advice is given.

Telephone-socket tester

This device is able to test for the on-hook DC line voltage and polarity between terminals 2 and 5 of a telephone socket, and provision is made for the test to be carried out on systems operating at a nominal 48 V, and those operating at around 30 V. In a separate test, the presence of the DC loop current is indicated, in the 'off-hook' condition, and the integrity of the bell-wire connection to terminal 3 is checked.

It will be seen from the circuit diagram (Fig. A4.1) that the circuit is arranged in three separate sections. The first section, to the left of the diagram, is permanently connected between terminals 2 and 5. Assuming that 48 V DC is present (negative on 2, positive on 5), current will flow through R1, D2, LED2, ZD4, and ZD2, and the green LED will be illuminated. If the polarity is incorrect, the current flow will be via ZD1, ZD3, LED1, D1, and R1, and the red LED will be lit. The sum of the zener voltages (44 V) will be subtracted from the 48 V line, leaving just 4 V to be dropped across the LED, diode and R1. Thus, if the voltage falls below about 45 V, the LED will not light.

For testing lower-voltage systems, the push-button switch S1 is momentarily closed, and the current flow for the correct polarity will be R1, D2,

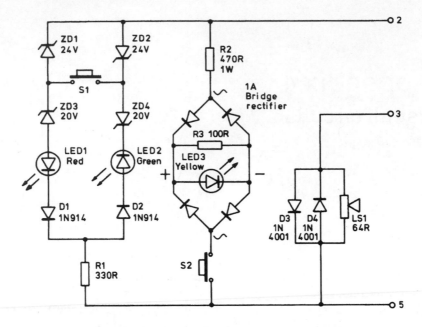

Figure A4.1   *Socket tester – circuit diagram.*

LED2, ZD4, S1, and ZD1. In this case, ZD1 is conducting in the forward direction and drops less than 1 V. For the incorrect polarity, a similar situation exists for the other leg of the circuit.

The remainder of the circuitry comes into operation when the pushbutton S2 has been pressed. Line current flows through the switch, through the lower left diode of the bridge rectifier, LED3, the upper right diode and R2. If the polarity is incorrect, the current flow is through the other two diodes of the bridge rectifier. As the loop current may be rather high for the LED, some of the current is shunted by R3.

While S2 is pressed and loop current is flowing, the exchange will respond by applying dialling tone to the line. This signal will pass through the capacitor connected between terminals 2 and 3 in the master socket, even though its normal function is to pass the ringing signal. The dialling, tone therefore, may be monitored with a suitable audio transducer connected between terminals 3 and 5. Note that the 64 Ω impedance of this miniature loudspeaker is considerably higher than the more usual 8 Ω type. The two diodes, D3 and D4, protect the speaker in the event of a ringing signal being applied to the line while the unit is plugged in.

A suggested method of constructing the unit is shown in Fig. A4.2, using a readily available pocket-sized plastic box into which the 64-mm-diameter speaker can be neatly fitted. A rocker-style switch is suggested for S2,

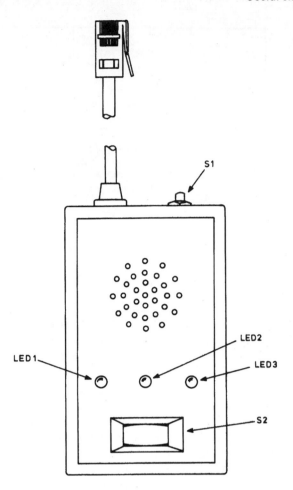

Figure A4.2 *Socket tester – suggested construction.*

while S1 is a miniature pushbutton mounted on the end of the case to avoid accidental operation.

Line-use detector (Fig. A4.3)

This device is designed to be connected permanently across a telephone line. The indicator LED is lit whenever a telephone connected to any socket on the same circuit is off-hook. Its purpose is to prevent accidental 'butting in' to a conversation already in progress.

If the telephone line voltage is greater than about 20 V, transistor Q1 will be turned on, and Q2 will be off so that no current flows through the

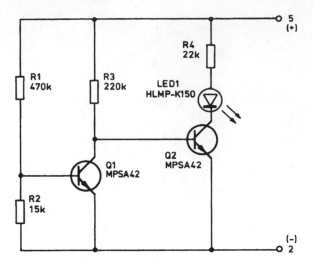

Figure A4.3   *Line use detector – circuit diagram.*

LED. When the line voltage falls below this threshold, indicating the off-hook condition, Q1 turns off, and Q2 conducts though the LED. The LED specified is a low-current, high-efficiency type, to ensure that the current drawn by the circuit is sufficiently low that the circuit conditions are unaffected and, more importantly, that the circuit does not latch on, holding the line engaged permanently. The circuit is polarity conscious and, as represented here, requires that the line polarity is correct. Do not be tempted to incorporate a bridge rectifier to accommodate either polarity, because modern exchanges such as System X apply a short, low-voltage line reversal on a periodic basis as part of an automated testing sequence. If the line-use detector reacts to this test, the exchange will register a fault condition.

The circuit has very few components, and because it is completely line-powered, it may be housed within a socket without difficulty. The specified transistors are designed to withstand the high voltages that are applied by the ringing signal, and low-voltage types should not be used.

Ring detector

The circuit operates a relay in synchronism with an incoming ringing signal, so that non-compatible alert devices may be used, or it could even be adapted for remote-control purposes.

At the heart of the circuit is an opto-isolator PC1 (see Fig. A4.4) that transmits a signal by optical means, thus providing complete electrical isolation from the telephone line. The ringing signal is arranged to turn on the

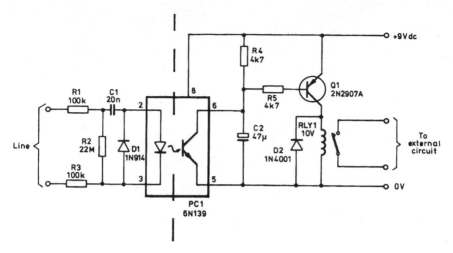

Figure A4.4 *Ring detector – circuit diagram.*

LED which is an integral part of PC1, which then turns on the photo-transistor. (Actually, the internal circuitry is slightly more complex than this to improve performance.) The output of PC1 turns on and off at the ringing frequency of about 20 Hz, in bursts corresponding with the ringing cadence. If this signal were to be used to drive the relay directly, it could 'chatter' instead of switching smoothly in time with the cadence. Components R4, R5 and C2 are therefore introduced to filter out the 20 Hz component, and the resulting signal used to operate the relay via Q1. The relay should be selected so that the contacts are rated appropriately for the intended purpose. Note the inclusion of D2, which is a 'snubber', to prevent high voltages generated within the relay coil during switching from damaging the transistor.

Caution: An external power supply is required to operate this circuit. The greatest care must be exercised to ensure that this power source, and any power source being switched by the relay contacts, cannot contact the line side of the circuitry, i.e. to the left of the dotted line shown in Fig. A4.4.

Two-station intercom (Fig. A4.5)

Many pulse-dialling telephones are now being abandoned in favour of tone-dialling models, and some are discarded because of dialling or ringing faults. With just a few components, a high-quality, two-station intercom can be constructed, which may make use of two of these redundant telephones.

There is no ringing generator, as this would be costly. Instead, low-cost audible warning devices such as the RS Components 245-001 are used to

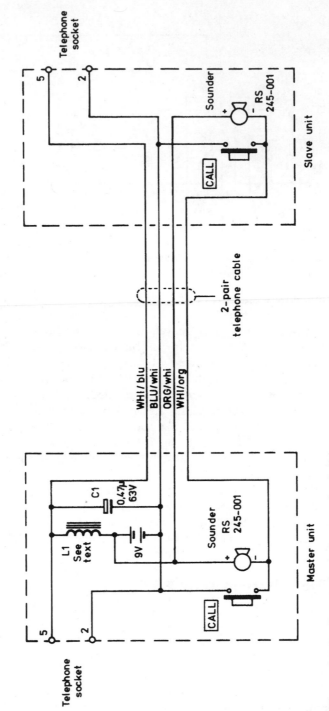

Figure A4.5 *Two-station intercom – circuit diagram.*

generate the calling signal. When either of the call buttons is pressed, both sounders will operate, providing user confidence. Line current is provided by a PP3-style battery, and no current is drawn when both telephones are 'on-hook'. The inductor, while important, is not critical. Its purpose is to prevent the audio signals on the line from being shorted out by the low effective resistance of the battery. The primary side of almost any small audio transformer may be used, such as the LT44 driver transformer (available from Maplin Electronics – see Appendix 3). If a suitable inductor is not available, a low-value resistor, in the range 50–100 Ω, may be substituted, but at the expense of much reduced volume and intolerance to a failing battery.

The capacitor across the line is included to perform two functions. Firstly, telephones are designed to work with much longer, and therefore more lossy, lines than might be expected for an intercom. The volume will be rather too high for comfortable use, and the capacitor will provide a degree of attenuation. Secondly, by using a capacitor rather than a simple resistor, the higher frequencies are attenuated more that the lower frequencies, and some of the fidelity of the transmitted speech is restored.

There is sufficient space to fit all of the components, including the battery at the 'master' end, into a 25-mm-deep back-box behind an LJU3/3A socket.

Index